D1346493

Bioelectronics

The Biotechnology Series

This series is designed to give undergraduates, graduates and practising scientists access to the many related disciplines in this fast developing area. It provides understanding both of the basic principles and of the industrial applications of biotechnology. By covering individual subjects in separate volumes a thorough and straightforward introduction to each field is provided for people of differing backgrounds.

Titles in the Series

Biotechnology: The Biological Principles: M.D. Trevan, S. Boffey, K.H. Goulding and P. Stanbury
Fermentation Kinetics and Modelling: C.G. Sinclair and B. Kristiansen (Ed. J.D. Bu'Lock)
Enzyme Technology: P. Gacesa and J. Hubble
Animal Cell Technology: M. Butler
Fermentation Biotechnology: O.P. Ward
Genetic Transformation in Plants: R. Walden
Plant Biotechnology in Agriculture: K. Lindsey and M.G.K. Jones
Biosensors: E. Hall
Biotechnology of Biomass Conversion: M. Wayman and S. Parekh
Process Engineering in Biotechnology: A.T. Jackson
Biotechnology in the Food Industry: M.P. Tombs
Plant Cell and Tissue Culture: A. Stafford and G. Warren (eds)
Bioprocessing: O.P. Ward
Bioelectronics: S. Bone and B. Zaba

Series Editors

Series Advisers

The Institute of Biology IOB

*This series has been editorially approved by the **Institute of Biology** in London. The Institute is the professional body representing biologists in the UK. It sets standards, promotes education and training, conducts examinations, organizes local and national meetings, and publishes the journals **Biologist** and **Journal of Biological Education**.*

For details about Institute membership write to : Institute of Biology, 20 Queensberry Place, London SW7 2DZ.

Bioelectronics

Stephen Bone and Bogumil Zaba

JOHN WILEY & SONS

Chichester · New York · Brisbane · Toronto · Singapore

First published 1992 by John Wiley & Sons Ltd
Baffins Lane, Chichester
West Sussex PO19 1UD, England

Other Wiley Editorial Offices

John Wiley & Sons, Inc., 605 Third Avenue,
New York, NY 10158-0012, USA

Jacaranda Wiley Ltd, G.P.O. Box 859, Brisbane,
Queensland 4001, Australia

John Wiley & Sons (Canada) Ltd, 22 Worcester Road,
Rexdale, Ontario M9W 1L1, Canada

John Wiley & Sons (SEA) Pte Ltd, 37 Jalan Pemimpin 05–04,
Block B, Union Industrial Building, Singapore 2057

British Library Cataloguing in Publication Data

Bone, Stephen
Bioelectronics. – (Biotechnology series)
I. Title II. Zaba, Bogumil III. Series
574.19

ISBN 0 471 93296 5

Typeset by Vision Typesetting, Manchester
Printed in Great Britain by
Biddles Ltd, Guildford

Contents

Chapter 1

Introduction

Birth of a discipline

Any new branch of science faces problems in defining the boundaries of its legitimate activity. Since new fields most commonly arise from the fusion of other, more traditional areas, the problem is often one of attempting to justify some new 'ABism' or 'XYology' in the face of criticism from long-standing X-icists and Y-ologists who claim that XYology represents at best a minor shift in emphasis in a field which is the domain of the traditional practitioners and at worst nothing more than a fancy name invented in order to obtain a research grant.

Nonetheless, the word 'Bioelectronics', although absent from the spell-checker of most word processors has come into increasingly common usage among biochemists, physicists and electronic engineers in the last decade. Probably the term 'bioelectronics' was first coined by the visionary biologist Albert Szent-Györgyi. In his two short volumes, *Introduction to a Submolecular Biology* (1960) and *Bioelectronics* (1968), Szent-Györgyi depicts a vision in which biology, once the study of whole living organisms and their interactions with their environment, becomes a science of the sub-molecular. The properties of living organisms become explicable in terms of the electrons which are responsible, ultimately, for the chemical reactions which 'drive' the organism. There can be no doubting the sheer scientific courage and ambition which such an aim entails.

There has, of course, always been a desire to understand living systems at a level of organization ever more subtle and microscopic. Thus, at the start of this century the realization that biochemical reactions were subject to the same laws of chemistry as other types of chemical reactions gave place to the still burgeoning science of biochemistry. It is not surprising therefore that Szent-Györgyi should, some fifty years later, reach down to the next level of organization, at which study had previously not been possible, i.e. at the sub-molecular, electronic level. This can be viewed as a perfectly logical development of the approach of biochemists and biophysicists throughout the 20th century.

There is however another strand which can be recognized in the development of bioelectronics. That strand was related to the realization that there were electronic properties of biological materials that appeared unique to electronic materials scientists who had hitherto concentrated largely on simple inorganic crystalline materials, with the occasional foray into organic substances and particularly organic polymers for dielectric materials. This involvement of electronic engineers and physicists was motivated both by a natural curiosity in the properties of another diverse class of materials and also by a perceived possibility that out there in the biological world were materials that might be useful to the engineer, if not here and now then at some point in the future. Allied to the interest in biological materials (biological 'hardware') as possibly useful components for exploitation in future electronic devices, there is an interest in the manner in which biological systems organize information processing ('biological software').

We may therefore discern two strands in the development of this sub-discipline: we may refer to them as the fundamental and the practical. This suggests immediately a good pedigree, since to flourish, any area of science needs both its 'internal' enthusiasts and their more technologically inclined cousins. The latter approach provides its own justification in terms of potential beneficial applications. It may be instructive however to ask ourselves exactly what technological benefits are envisaged.

The technological challenge

What is it that biology can teach the electronic engineer, or what does biology do better than man-made electronic devices when the two types of system are in competition?

The challenging aspects of biology from a modern electronic engineer's viewpoint are threefold:

(1) Biological systems store and process information using individual molecules, or sub-components of molecules as the storage medium. We may refer to this phenomenon as *biological information processing*.
(2) Biological systems are able to direct electronic (and ionic) currents along pathways that are also defined at the ultramicroscopic (molecular) level. This we may refer to as *biological microelectronics*.
(3) Biological systems have developed a range of exquisitely sensitive transducing systems for receiving information about the state of their environment. These are *biological microsensors*.

Finally, we need to realize that all of these biological molecular devices share a feature which will be desirable if not essential in any future man-made molecular device – they are essentially self-assembling. That is to say, it is sufficient for the components of the device to come together (sometimes in the presence of an appropriate catalyst) for the device to 'self-construct'. This is achieved by the sub-components having surface sites which have a high affinity for complementary sites on the other sub-components.

BIOLOGICAL INFORMATION PROCESSING

The first of these achievements of the biological world appears to have little to do with 'electronics' *per se* since the reactions involved in storing and processing biological information (such as the replication of nucleic acids) are usually thought of as 'ordinary' chemical reactions, involving no charge transfer. There is no doubt, however, that the net result of these reactions involves the storage of vast amounts of information in extremely compact structures.

A rough calculation might be instructive here. DNA molecules, such as those found in human chromosomes store roughly 4 Gbits of information in a length of some 900 mm. Knowing the diameter of the DNA helix (2.2 nm) we can calculate the information storage density. This works out at roughly 1.2 Gbits/μm^3. If we undertake a similar calculation on the information storage density of a modern microcomputer magnetic (hard) disk, we find that the equivalent figure is 4×10^{-12} Gbits/μm^3. This suggests that the molecular technique has an advantage over the magnetic technique of some 12 orders of magnitude. Optical storage technology (CD-ROM disks), which represents our most advanced current technique for information storage, give an advantage for some two orders of magnitude over magnetic media.

Biological miniaturization does not end here, however, with the information storage medium, but it applies equally to the mechanisms required to manipulate that information. So all the machinery required to copy the genetic information is contained in the cell nucleus, a sphere with a diameter, in higher organisms, of approximately 20 μm. That is to say the functional unit, which is capable of copying and producing selective transcripts for export (which we might think of as equivalent to disk copy and file copy operations in a digital computer), has a size of 33 500 μm^3, or 33×10^{-6} mm^3. A disk drive unit, which is capable of similar operations is of course a large structure, some 1000 cm^3, or 1×10^6 mm^3 in volume. Again we see 12 orders of magnitude advantage in size for the molecular machine.

If the performance of the hardware in biological information processing is impressive, then so too is that of the software. In the case of DNA, of course, we have contained within the genetic material not only the code for all the components of an organism, but also the code for all the necessary machinery to construct that organism. Extensive error-correcting mechanisms are built into the replication machinery. Biological software is particularly impressive in areas where much parallel processing of input signals is necessary, for example in image and pattern processing. Recent work suggests for example that parallel paths exist which process information on colour and motion when the eye is analysing a moving object. Pattern recognition and matching imperfect copies of visual and aural information remains one of nature's information processing tricks which current computing technology simply cannot match. There seems little doubt that there are important lessons to be learnt in both nature's algorithms and information processing architectures. Recognition of this fact underlies the current rapid expansion in research in neural networks, where the parallel processing capabilities of neural systems are being emulated in both hardware and software.

BIOLOGICAL MICROELECTRONICS

When it comes to biological microelectronics, as opposed to biological information processing, another set of fascinating phenomena are revealed. Biology, it seems, also has solid-state devices. But whereas the man-made variety rely on bulk properties of layers of conductors, semiconductor and insulators to control and confine the flow of electronic charges, the biological devices are assembled from individual electroactive macromolecules. These have the necessary properties to either confine or transfer electrons or ions usually by providing a binding site (for electrons an electroactive centre – typically a transition metal ion). The affinity of the binding site can often be manipulated depending on the interaction of other (effector) molecules with the macromolecules containing the electroactive centre. As will become apparent in Chapter 6, a major area of current research is in the elucidation of how these electron-transferring macromolecules are arranged spatially with respect to one another. Do they occupy fixed positions, or are they able to move freely with respect to one another?

It is worth mentioning speed here, since this is a subject which is inevitably raised when comparisons are being made between man-made and biological devices. Attention is often drawn to the fact that switching times in man-made devices are several orders of magnitude shorter than those found in biological devices. There is no doubt that this is true, but set against this we must be mindful of the enormous scope for miniaturization that a truly molecular architecture promises. Particular attention is, of course, being focussed on the possibility of imitating the molecular microstructure and self-assembly of the biological systems but using the switching mechanisms of the man-made electronic components.

BIOLOGICAL MICROSENSORS

It is almost a truism to say that biological systems are extremely sensitive to their surroundings and are able to detect very subtle changes in the environment. It is only recently, however, that it has become apparent that the sensitivity limits of a number of these systems are quite close to theoretically limiting values. Thus the human eye (and that of many other vertebrates) can function as a single photon counter. That is to say, the signal arising from the interaction of a single photon with the retina is under appropriate conditions readily distinguishable from the background noise in the system.

Similar sensitivities, approaching theoretical fundamental limitations in signal-to-noise ratios apply, for example, to sound transduction in the mammalian ear and to various forms of chemoreception. The latter phenomenon, perfected in even the most primitive unicellular organisms allows not only the detection of extremely low concentrations of specific molecules (with an absolute limit of ca. 50 molecules), but also the detection of very shallow concentration gradients existing at these low absolute levels. Behavioural responses (usually movement of the organism in the direction of increasing concentration of food supply/potential mate) follow the detection of such gradients. Finally, we have some fascinating examples of sensitivity to electromagnetic fields. It has been known for some time that certain species of fish use weak electromagnetic fields for navigational purposes. It has recently been confirmed that some mammals also have this facility. The duck-billed platypus, it appears has electroreceptors which are able to detect electric fields of a strength as low as 60 μ V/cm. The detection of these fields allows the animal to stalk its prey since the muscular activity of the shrimp gives rise to fields of this strength.

The intellectual challenge

The other approach to bioelectronics, based on curiosity and a desire to provide descriptions of living systems at ever-more fundamental levels of organization, also requires some scrutiny. It is fair to ask two questions:

- Will it be useful and relevant to have a description of living systems at a level so far divorced from the final form of living matter?
- Is it realistic to try and produce such a description?

To answer the first of these questions we have to investigate the adequacy of our present physico-chemical descriptions of biochemical systems. An apparently non-electronic, purely biochemical topic such as drug evaluation and design might be instructive. There is no doubt that in many instances classical biochemical approaches provide a reasonable basis for prediction of behaviour of the systems of interest in the short term. For example, we know enough about the enzyme systems (and their control) to predict the fate of major metabolites in cellular degradative pathways. In many instances we also are able to predict how this physico-chemical behaviour will change in the presence of a variety of natural and man-made agents (drugs). Our predictions tend to be based, however, on a knowledge base of how similar compounds have been seen to interact with such systems in past experiments, rather than on any models of direct interaction of the given compound with components of our metabolic system. This has proved to be a handicap when attempting to rationally design novel therapeutic agents. The situation is changing dramatically at present with the advent of molecular modelling techniques. Among the fundamental input parameters of molecular modelling, however, one has electronic charge density functions. In order to be able to predict how a new drug will interact with an enzyme in a metabolic pathway it is necessary therefore to understand something of the interactions between the electronic charges carried by the two molecules – that is, one is pushed into the field of bioelectronics. While not every practitioner of biochemical molecular modelling would think of themselves as working in the field of bioelectronics, there is little doubt that the tools of their trade are precisely the type of tools which Szent-Györgyi had in mind when he said:

> Looked at through the glasses of this new science [quantum mechanics] the atom is no more an invisible unit but consists of a nucleus surrounded by a cloud of electrons with varying and fantastic shapes, and it seems likely that the subtler phenomena of life consist of the changing shapes and distributions of these clouds.... What admits no doubt in my mind is that the Creator must have known a great deal of wave mechanics and solid state physics, and must have applied them. Certainly, He did not limit himself to the molecular level when shaping life just to make it simpler for the biochemist.

To broaden the argument somewhat, one of the unique features of biochemical as opposed to chemical systems is the high degree of specificity with which reactants are brought together on the surfaces of the catalysts. One thinks, for example, of the many biological catalysts which distinguish clearly between optical isomers of both simple and complex molecules or the exquisite specificity which antibodies secreted into the bloodstream seek out and bind to their corresponding antigen. Antigen–antibody interaction has been the subject of much detailed investigation in recent years. For a number of specific cases, we know in some detail the amino acid residues involved in the binding process. For the case of lysozyme binding to its antibody for example, an area of some 20 × 30 Å is involved in which some 16 different amino acids from the lysozyme molecule make hydrogen or van der Waals bonds with corresponding residues on the antibody surface.

The interactions which give rise to this specificity are predominantly electrical. While quasi-mechanical descriptions (locks and keys, molecular levers, etc.) are useful in visualizing these processes, predictions require a greater level of sophistication. It is in order to make useful predictions of biomolecular interactions that one is forced to study the relevant biomolecules at ever lower levels of structure and organization including the electronic level.

When it comes to the second question posed above – how realistic is it to look for an electronic description of biomolecules? – we must attempt an honest assessment of our present and likely future capabilities. There is little doubt that a complete description, in terms of a model with no adjustable parameters, which will successfully predict the behaviour of even the smallest of proteins in all situations on the basis of its known chemical composition and hence electronic structure, is a very long way away from current capabilities. It is instructive in this context to survey the success of chemists in describing relatively simple molecules on the basis of first principles. An obvious starting point is the prediction of electronic spectra on the basis of the known electronic structure of a chemical entity. A typical text in this field will contain chapters on atomic spectra, proceeding to diatomic molecules and finally to polyatomic molecules. While the first chapter will be rigorous, and will use the formalism of quantum mechanics, that is, the best description we have currently available to describe and predict the behaviour of electrons, the subsequent chapters will deviate more and more from the rigorous and will contain more empirical constants, assumptions and approximations. These are required to deal even with such relatively simple molecules as aromatic hydrocarbons containing some tens of atoms. We must therefore, however reluctantly admit that the prediction of the behaviour of even the smallest protein, containing as it does some thousands of atoms, from a consideration of its electronic structure is a long, long way beyond our current capabilities.

A recent review of the photosynthetic reaction centres suggested that to compute the wavefunctions for the various states of this protein complex would be some ten orders of magnitude beyond present-day computing facilities. We are therefore in the somewhat frustrating position of seeing a considerable potential for a new description of biomolecules but having to admit an inability to grapple with the problem due to the complexity of the systems. How then should the bioelectronic approach be applied?

Realistic goals

Our approach in this book will be a pragmatic one, that is, to apply the bioelectronic approach to areas in which it seems most likely to yield useful results. Where our understanding of biological materials appears to be enhanced by applying electronic principles then those principles should be applied. Where the exercise appears to be altogether too ambitious, we shall not attempt it. Similarly, where technological potential appears to be realizable through a bioelectronic approach, again it seems obvious that the attempt should be made. Since artificial molecular information processing, while clearly an important long-term goal is still over the horizon, we will not concern ourselves further with this area. Our pragmatic approach will lead us into four main areas, outlined below.

BIOLOGICAL ELECTRON-TRANSFER REACTIONS

Here we are clearly dealing with electrons and it would be surprising if the progress in understanding electronic behaviour in physico-chemical terms did not have something to contribute to the traditional biochemists' approach.

An example of the type of system which has been of interest to biochemists for many years is shown in Fig. 1.1. The schematic diagram of one of the biological electron-transfer 'chains' illustrates a number of results which have come from a classical biochemical approach, viz. fractionation of a complex system into constituent parts, followed by analysis of those parts and finally reconstitution of the parts into a whole. But we are also aware, when looking at such a diagram of our ignorance of many aspects which would allow us to truly understand how such electron-transferring chains work. How are the large complexes organized with respect to one another? What are the factors which limit the flow of current through the system? How precisely is a proportion of the energy of the electron, which is surrendered as it travels from low to high redox potential trapped to generate reactive chemical intermediates? These are all questions which require an approach more subtle than the biochemist's standard destruction and reconstruction technique.

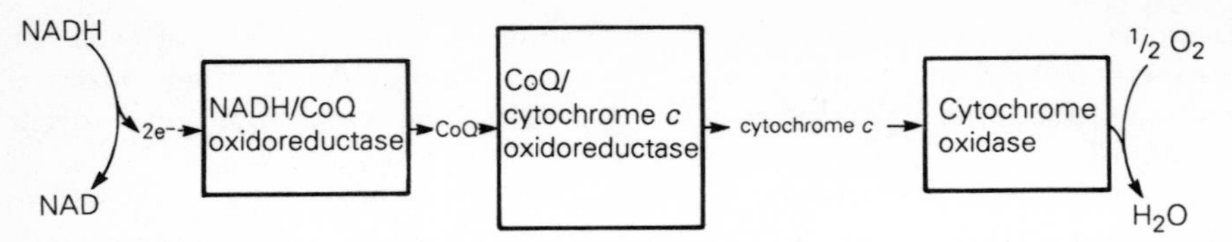

Fig. 1.1 The mitchondrial electron-transport chain. The boxed entities are complexes containing multiple electroactive proteins. Between the complexes electrons are carried by smaller, more freely-diffusing molecules.

ELECTROCHEMISTRY INVOLVING BIOLOGICAL MATERIALS

Here we have an example of an area in which potential practical applications are envisaged in the near term, in the form of biosensors. There is also every hope that in designing such sensing devices, we will come to better understand how to interface biological molecular materials with conventional electronic circuitry, thus opening the door to more ambitious projects involving molecular computation. Various types of interface between biological materials and electronic components which are the subject of current study are illustrated in Table 1.1.

DIELECTRIC PROPERTIES OF BIOLOGICAL MATERIALS

Here is another instance where a highly successful formalism exists which has been applied to relate bulk properties of materials to their atomic and electronic structure. As will become obvious, we can enhance our understanding of the structure and function of biological materials by applying this formalism to biopolymers.

WATER IN BIOLOGICAL SYSTEMS

As soon as we attempt the exercise referred to above, we become aware of the very important role of water in biological systems. Much of the dielectric responses

Table 1.1 Methods used to 'communicate' with active biomolecules

1 Electrochemical
- Direct
- Mediated
- Surface-modified electrode

2 Dielectric
- Frequency domain
- Time domain

3 Optical
- Absorption
- Fluorescence
- Surface plasmon resonance

found in biological materials can be attributed to water molecules, both freely mobile in the bulk solvent, and, more interestingly, bound to various degrees to the biological materials in question.

Selected reading

Nicholls, D.G. (1982). *Bioenergetics*. London: Academic Press.

Szent-Györgyi, A. (1960). *Introduction to a Submolecular Biology*. New York: Academic Press.

Szent-Györgyi, A. (1968). *Bioelectronics*. New York: Academic Press.

Turner, A.P.F., Karube, I. and Wilson, G.S. (1987). *Biosensors: Fundamentals and Applications*. Oxford: Oxford University Press.

Chapter 2

Principles of Bioenergetics and Bioelectrochemistry

Introduction – biological charge-transfer systems *in vivo* and *in vitro*

In this chapter we will be concerned with introducing biomolecular micro-charge transferring devices. As will become clear, these devices consist of protein molecules and supramolecular complexes of these molecules in which the molecular organization provides pathways for electrons, protons and other ions to move. The charge-transfer process can be from high to low potential or vice versa. In the latter case, energy must clearly be provided and a primary function of these devices in the living cell is in interconverting chemical energy, as stored in the bonds of organic molecules and electrochemical energy, as charge-separated systems (e.g. electrochemical gradients) and vice versa. Additionally, we have the closely-related photosynthetic systems which transform electromagnetic radiation to electrochemical gradients and hence to chemical bond energy. Before we set out on a detailed exploration of these systems however, it will be useful to outline some general properties of biological macromolecules and to take note of some of the problems which face those who wish to exploit the undoubted potential of these systems in novel man-made device structures.

Any elementary course in biological chemistry begins by classifying the chemicals of life into a small number of groups. These are the lipids, the carbohydrates, and nucleic acids and the proteins. These classes can, at the risk of considerable over-simplification be associated with the functions of compartmentalization, energy supply and storage, information storage and retrieval, and catalysis respectively. In addition, all of these types of materials with the exception of the nucleic acids, fulfil structural roles in living systems. We will be concerned with only two of the above classes of materials – the lipids and the proteins. Only a very brief introduction can be given here. Readers are referred to a number of standard works for a more detailed discussion.

The lipids which we will be concerned with will be mainly the phospholipids which make up the membranes of living systems. Unlike most other classes of biological materials, the phospholipids are insoluble in water. They are in fact amphiphiles, having both hydrophilic and hydrophobic elements in their structure. The result of this structural property is that in water and aqueous solutions of simple salts, phospholipids spontaneously form bimolecular bilayer structures in which the hydrophobic hydrocarbon 'tails' come together, away from the solvent, leaving the hydrophilic phosphate ester groups in contact with the solvent. Phospholipids can therefore form sheets of ultra-thin bilayer membrane, and living systems make use of these structures to form membranes surrounding cells and also, in the case of higher organisms to provide compartmentation within cells. The membranes thus formed are electrically highly insulating with conductances of $<10^{-4}$ San^{-2}.

The proteins provide the most structurally and functionally diverse group of biomolecules. As heteropolymers of twenty different amino acids (a chemically diverse group themselves), with chain lengths of 100–400 monomer units, their richness of structural possibilities can be imagined. Among the catalytic proteins – enzymes – we will be primarily concerned with two sub-groups: soluble enzymes and membrane-associated enzymes. The former can be found dissolved in the internal cellular electrolyte. The structure of these proteins also makes use of the hydrophobic effect. In the folding of the polymer chain, hydrophobic amino acid residues are internalized, while hydrophilic residues are on the outer surface in contact with water. Hydrogen bonding between the externalized amino acids and the solvent water stabilizes the structure.

The membrane-associated proteins, however, show very different structural features with hydrophobic amino acid residues on part of the exposed surfaces. These surface regions with hydrophobic properties interact with the hydrophobic 'core' of the phospholipid membrane to anchor these proteins within the membrane structure. In addition to these hydrophobic 'domains', membrane proteins typically have one or more regions of hydrophilic surface, which will be in contact with the aqueous medium on one or both sides of the membrane. This varied protein–lipid interaction gives rise to a 'fluid-mosaic' membrane, a term first used by Singer and Nicholson in 1972. In this mosaic, both the phospholipids and proteins are considered to be mobile in the lateral plane, although there is little 'flip-flop' movement across the bilayer.

The proteins within the insulating membrane can themselves be electronically or ionically conducting. They can also assemble, with other proteins, into complexes with conducting properties. In general, however the conduction is highly anisotropic. That is, there are specific pathways, defined at the molecular level, along which electrons or ions can move from one part of the molecule or complex to the other.

Electronic conduction is usually achieved through the use of redox centres which are attached to the protein. Conduction is by hopping or tunnelling from centre to centre. Pathways may be either transverse or lateral to the plane of the membrane. Of interest is the fact that the rate of electron transport (current) can be modulated by many chemical and physical factors which change the relationship of one redox centre to another.

Ionic conductivity arises from two rather different systems. They are usually referred to as *pores* and *pumps*. The pore-forming proteins which are present in many membranes act to allow the passage of ions through aqueous or quasi-aqueous channels. Although these channels may be 'gated', i.e. electrical or chemical events may open or close them, a characteristic of their function is that ions may only move through them *down* their electrochemical gradient. The pores or channels therefore act to facilitate a process which is thermodynamically spontaneous. In contrast to this mode of ion movement, there are the ion 'pumps'. These systems are more complicated than the pores and are able to move ions *against* prevailing electrochemical gradients. Energy required for this endergonic process is derived from coupled exergonic processes (such as ATP hydrolysis).

It is tempting at this point to begin to speculate on the various devices which could be built out of the electroactive building blocks described above. After all, we have molecular conductors (sequences of electron carriers), molecular batteries (chemically-powered ion pumps) and molecular transistors (chemical modulation of current). It is important however to inject a note of caution which will be applicable to all discussion of exploitation of biological materials in future devices. This relates to the problem of the stability of the active materials. Biological materials (and here we are dealing with proteins for the most part) are designed to play highly specific and often remarkable roles in the molecular machine that is the living cell. However, they are designed to function within the protected and highly controlled environment of that cell. When we remove them from the cell and purify them in order to extract a molecule with some useful function, we cannot be sure that it will behave in the hostile environment of the test tube in the same way as it behaved in the cell, surrounded as it was by often precisely aligned neighbouring proteins, lipids and carbohydrates.

The biochemist has long been faced with a dilemma – whether to purify cellular components to homogeneity (and possible totally artefactual behaviour) or to attempt the difficult and complicated task of studying all components of interest in the complex original cellular environment. Inevitably, both courses of action have been pursued, and from correlations found between results obtained from the two approaches, important clues have been obtained as to the true nature and function of the components. The favoured approach has often been to purify the protein, characterize it as a pure substance and then to re-incorporate it into some structure mimicking its original environment. Such reconstitution studies have been carried out with many of the interesting electron-transport proteins which can be purified from membranes and then reconstituted into artificial membranes of various types. It is however a fact of life with which the protein chemist has to learn to live, that purified proteins do not have an indefinite shelf-life, and that high temperature, extremes of pH, hydrolytic enzymes (present in many crude cell extracts), metal ions and many other chemical and physical factors can adversely affect protein stability. For those unfamiliar with the techniques of protein

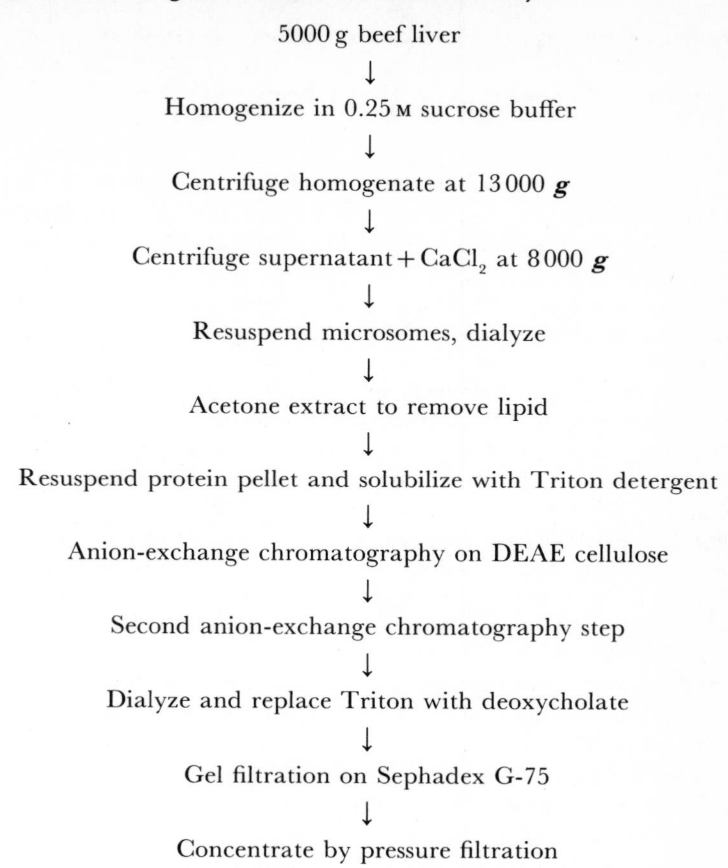

Fig. 2.1 Scheme for preparation of cytochrome b_5.

extraction and purification, it may be instructive to study Fig. 2.1, a scheme for the preparation of milligram quantities of a small, membrane-bound electron-transferring protein, cytochrome b_5. Clearly the production of such proteins in large quantities, in active form and high purity, is not a trivial undertaking.

We must therefore bear this point in mind. The mere identification of an exciting biological molecular function represents no guarantee that such a functionality can be successfully isolated and exploited. Typically, there will be a long and tortuous route from identification to exploitation, both in terms of finding a purification protocol which works, finding conditions for correct functioning and providing a functional interface with other components.

Bioenergetics – biological electron and proton transport

ELEMENTS OF BIOLOGICAL CHARGE-TRANSFER SYSTEMS

Electron-transfer systems
The obvious place in which to start the study of electron-transferring elements in biological systems is the so-called *electron-transfer chain*, present in the membranes of mitochondria in higher organisms and in the cell membranes of lower unicellular organisms such as the bacteria.

The metabolism of foodstuffs such as glucose consists of a series of reactions, the aim of which is to oxidize the glucose (remove electrons) so that the energy change inherent in this (exergonic) process can be utilized for a series of useful purposes. The biological function of the electron-transport chains is to transfer reducing equivalents (electrons) from NADH to oxygen – the ultimate electron acceptor in aerobic oxidation processes. The chains consist for the most part of electroactive proteins plus a number of other low-molecular-weight electroactive components (e.g. the quinone-based coenzyme Q and its relatives). These chains are of fundamental importance in aerobic metabolism for two reasons: (i) They are the mechanism by which the redox cofactor NADH, which is reduced in the early stages of (anaerobic) glucose metabolism is reoxidized to NAD (and therefore 'recycled' back into metabolic activity); and (ii) the electron-transport chain makes available the energy inherent in the transfer of an electron from the NAD/NADH couple to the oxygen/water couple (220 kJ/mole or 1.14 V) for the synthesis of the cell's energy 'carrier', ATP. Three moles of ATP are synthesized for every mole of NADH oxidized. For this reason electron-transport chains of this type are known as *energy-conserving* or *energy transducing electron-transport chains*. Figure 2.2 attempts to place the electron-transport chain in its relevant metabolic

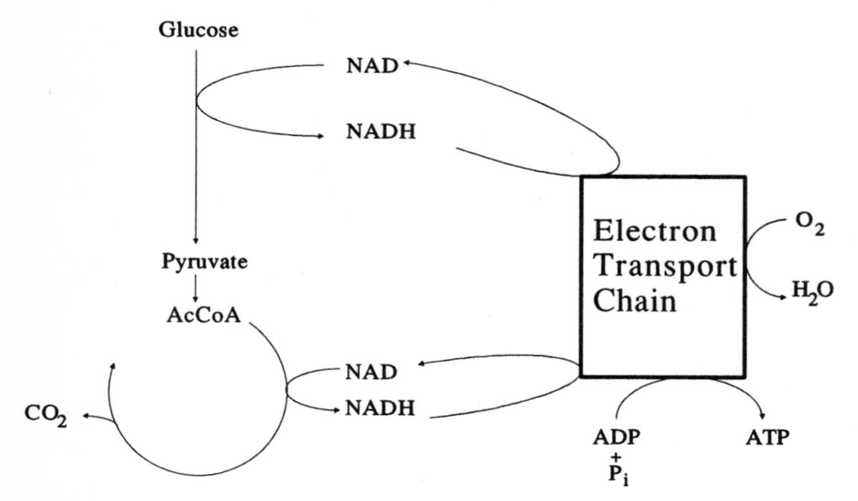

Fig. 2.2 The electron-transport chain in its metabolic context.

context. These chains are notable also for the controversy that they have generated among biochemists, who have been attempting to unravel the precise molecular mechanism by which they function as energy-transducing devices ever since their discovery in the mid-1950s.

For our present purposes, we may ignore this area of controversy, although we will come back to it many times later. If we ignore the energy-transducing aspect of these systems, we can concentrate on them as chains of electroactive biomolecules and we can attempt to categorize these components. Figure 2.3 shows the main components of the mitochondrial electron-transport chain. It should be noted that the chain contains a wide variety of carriers some of which carry a single electron, some an electron plus a proton and yet others multiple electrons and/or protons. Nonetheless, the chain is usually referred to as an electron-transport chain in order to underline the fact that the fundamental overall process which is catalyzed is a redox reaction with a number of intermediate steps – i.e. an electron-transfer from NADH to oxygen. As mentioned above, the electrons are carried by both macromolecular components (redox proteins) and small electroactive molecules.

An important feature of these chains which should be noted at the outset is that they are located in membrane structures made up of insulating lipids as described above. In the cells of simple unicellular organisms, the membrane containing the electron-transport chain is the membrane surrounding the whole cell. In the case of the more complex multicellular organisms, the relevant membrane is usually the

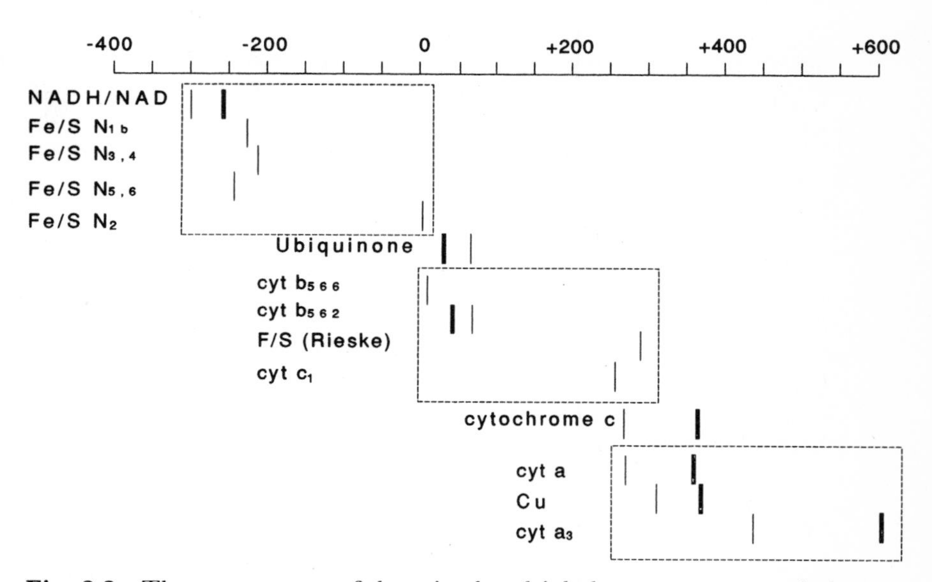

Fig. 2.3 The components of the mitochondrial electron-transport chain. Each component is mapped according to its redox potential. The thin lines show mid-point potentials while the heavy lines indicate values for the redox potential under energized conditions. Dotted boxes indicate components which are organized into supra-molecular complexes.

inner membrane of an organelle specialized for ATP synthesis – the mitochondrion.

Among the proteins, a number of different types can be identified: (i) haem proteins, such as the cytochromes; (ii) iron–sulphur proteins; (iii) flavoproteins. The first two categories are carriers of electrons only, whereas flavoproteins carry electrons plus protons (i.e. hydrogen atoms). The different classes of proteins are characterized by their different prosthetic groups, which are responsible for the observed redox activity. In each case, it is the prosthetic group of the protein which is actually reduced or oxidized in the course of the electron-transfer reaction.

Haem proteins In the haem protein, the prosthetic group is the haem moiety, consisting of a tetrapyrrole ring with an iron atom coordinated to the pyrrole nitrogens. It is the iron atom which can exist in both the Fe(II) and Fe(III) state which is the redox active component of the system. The haem proteins are in general one-electron transfer proteins, although complexes of proteins in which more than one haem group is present are known (e.g. cytochrome a/a_3, the terminal oxidase of the mitochondrial electron transport chain), and in this case clearly multiple-electron transfers to the protein complex are possible. The various haem proteins differ from one another in the manner in which the haem group is attached to the polypeptide chain(s) and in the nature and extent of substitution on the tetrapyrrole ring.

Iron–sulphur proteins The iron atoms in these proteins are found in close association with groups of sulphur atoms, which are usually attached to the protein backbone in cysteine residues. Iron–sulphur clusters usually contain equal numbers of Fe and S atoms, common combinations being [4Fe+4S], [3Fe+3S] and [2Fe+2S]. As well as containing multiple Fe atoms within the iron–sulphur group, many iron–sulphur proteins also contain multiple iron–sulphur groups. Although in many cases multiple oxidation states can be formulated for the iron–sulphur centres (e.g. $[4Fe+4S]^{3+}$, $[4Fe+4S]^{2+}$, $[4Fe+4S]^{+}$) each individual [4Fe+4S] centre often accepts only one electron in the course of its physiological redox reaction. The multiple iron–sulphur centres are thought to act to 'store' electrons in situations where the iron–sulphur protein acts as an intermediate between a one-electron redox system and a two-electron system – a situation which arises frequently in biological electron transport chains. In such a situation two Fe–S clusters can become sequentially reduced by the one-electron transferring component, then passing two electrons simultaneously at the same potential to the next (two-electron requiring) component in the chain. While this is an elegant hypothesis, at the same time it has to be admitted that there are many instances of multiple Fe–S clusters in large protein complexes, for which no such 'electron store' functions can be ascribed. Their precise role and mode of participation in the redox process remains something of a mystery.

Flavoproteins Flavoproteins are very widely distributed redox enzymes which are characterized by the presence of a tightly, but non-covalently bound flavin nucleotide. There are two types of flavin nucleotide – the mononucleotide (FMN) and the adenine dinucleotide. In each case it is the isoalloxazine ring structure which undergoes the reversible two hydrogen ($2e^- + 2H^+$) redox reaction. The flavoprotein redox enzymes can be broadly classified into metal-containing and metal-free categories. Where a metal is present it usually cooperates in some way with the flavin moiety in achieving the desired redox properties of a particular enzyme. The metal most frequently associated with flavoproteins is iron and it is most commonly present in the form of Fe–S clusters. This means that a large protein complex such as the first complex shown in Fig. 2.3 is actually both an iron–sulphur protein and a flavoprotein, with the Fe–S clusters and the FMN complexed to different polypeptide chains which make up the large complex.

Photosensitive reaction centres All of the components mentioned so far generally participate in electron-transfer reactions in which electrons are transferred down the prevailing thermodynamic gradient – i.e. from low redox potential centres to those having a high potential. There exist in biological systems another class of important electron-transferring proteins which are light sensitive and in which light energy is used to promote electrons from high to low redox potential. The details of the internal functioning of these systems are described in Chapter 6. For now we should just note that among the components involved are chlorophyll pigment molecules (also based on the tetrapyrrole structure) together with cytochromes and quinones all arranged in a very precise spatial relationship which appears to be designed to maximize the efficiency of charge separation following the primary photochemical event.

Ion-transfer systems

Most of the significant ion-transfer systems which have been fully characterized are transmembrane ion-transfer systems. These systems are commonly classified into two categories: (i) ion channels and (ii) ion pumps. The latter are in general more complex systems since they catalyze by definition an energy-requiring process – the 'pumping' of ions against the prevailing electrochemical gradient. Ion channels, by contrast, can be relatively simple enzymes which simply provide a more or less specific channel for an ion or group of ions which can use this channel to diffuse down their electrochemical gradient, although the fact that the activity of these channels must be regulated in some way increases their complexity. Examples of each of these two types of ion-transport system are discussed below.

Ion channels – the voltage-gated Na^+ channel The classical example of an ion channel is the Na^+ channel of nerve cells. The activity of this channel is responsible for the conduction of nervous impulses in the excitable cells of animals and as such plays a crucial role in all aspects of animal function, particularly advanced (locomotor and brain) functions. This protein is one of the most exhaustively studied of all membrane proteins. It has been thoroughly characterized as regards amino acid composition and sequence, has been purified to homogeneity and has been reconstituted into artificial membrane systems to allow its function and regulation to be studied. The regulation of this channel is of particular significance to students of bioelectronics since the ion current carried by the channel is regulated

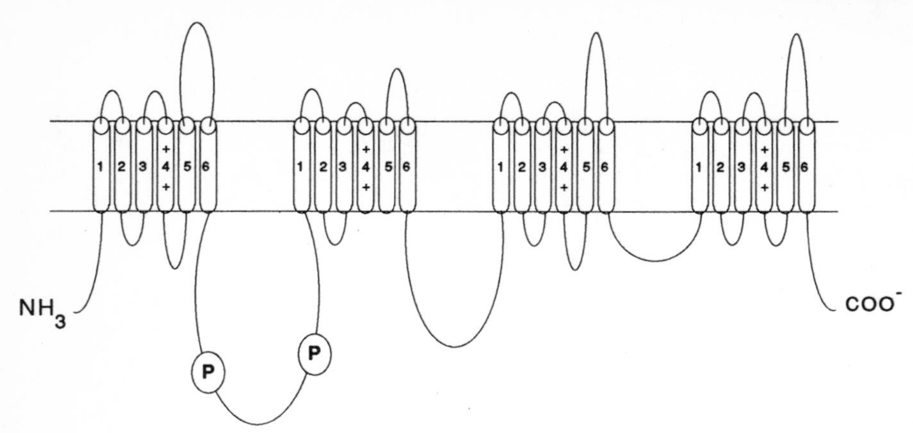

Fig. 2.4 Predicted structure for the voltage-gated sodium channel. P marks phosphorylation sites. Transmembrane helical segments are shown as numbered cylinders.

by the transmembrane voltage. The precise mechanism by which this electrical control of channel activity (known as 'voltage gating') is achieved is unknown despite the large amount of detailed information that has been gathered on this enzyme. As with many membrane proteins, the three-dimensional structure has not been precisely determined but has been partially predicted from the known amino acid sequence. The sequence upon which such predictions are based has features in common with a number of other voltage-gated channel proteins (e.g. K^+ and Ca^{2+} channels), enabling us to treat this channel as a representative member of a family of channel proteins. As these predicted structures become more detailed, and more information is added to them from chemical modification studies and immunochemical analysis, so models for the functioning of the proteins can be elaborated.

The predicted structures reveal four repeated structural domains each of which contains six hydrophobic helices, long enough to span a typical bilayer membrane. Figure 2.4 shows the resulting model. Recent work has led to the suggestion that one of the six helices (number 4) is a candidate for the voltage sensing element, since it is highly conserved between many different channel proteins and carries a characteristic pattern of positive charge (every third residue over a 20–30 residue length carries a positive charge). This leads to a less schematic model for the protein (Fig. 2.5) in which the protein is seen as forming an aqueous pore, with a voltage-sensing element controlling a 'gate'. The nature of this gate remains to be revealed. A further feature of this model is the selectivity filter, which determines the ion specificity of the channel. Clearly by altering the dimensions of, and charge configuration around, the 'filter', the protein structure can be adapted to act as a channel for a variety of ions.

It may be of some interest to compare the dimensions of this molecular switch with current electronic switches. As can be seen from Fig. 2.4, the whole structure

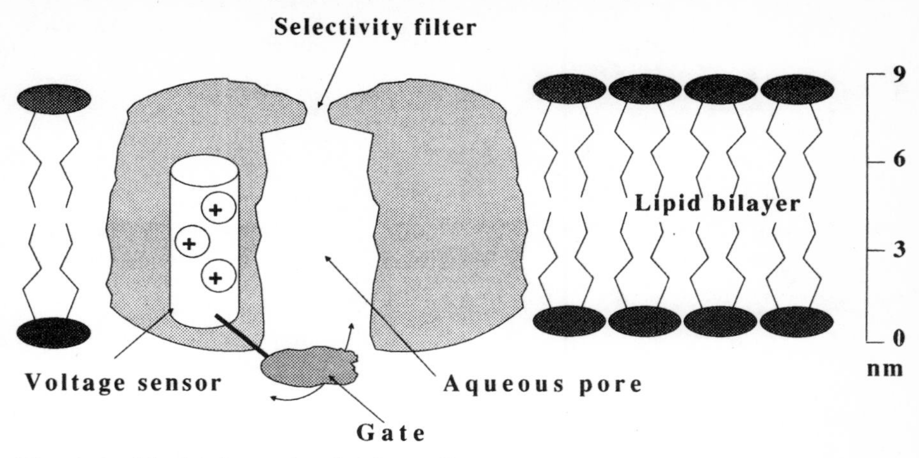

Fig. 2.5 Model for gating in the sodium channel.

has dimensions of the order of a few nanometres (3–5 nm). By comparison, today's state-of-the-art microprocessors contain hundreds of thousands of switches in a space of a few square centimetres, giving dimensions for the individual switches of the order of tens of micrometres. It is clear that in terms of sheer miniaturization, evolution has produced the more spectacular device by some three orders of magnitude.

Ion pumps – the Na^+*,*K^+*-ATPase* The majority of ion pumps use, as a source of free energy, the hydrolysis of ATP. Over recent years the ATP-powered ion pumps have been fruitfully studied by a barrage of physical, chemical and biological techniques which have resulted into their classification (on the basis of similarities in subunit structure and mechanism of action) into three main classes. These are F-type (mitochondrial and chloroplastic H^+-ATPases), V-type (plant vacuolar H^+ + ATPases) and the P-type ATPases which are responsible for a very wide range of ion-transport activities – Na^+, K^+, H^+ and others. The common structural feature is a 100-kDa catalytic subunit which becomes phosphorylated during the catalytic cycle, the phosphate group being transferred from ATP. The best-studied example of a P-type ATPase is probably the Na^+,K^+-ATPase. A simplified postulated reaction scheme for the enzyme is shown in Fig. 2.6. It is obvious that this is a highly evolved and sophisticated chemical device for charge transfer across an insulating membrane. As with the voltage-gated sodium channel, a model for the enzyme structure, based on the known amino acid sequence can be elaborated. The various P-type ATPases appear to share many common structural features.

Again, a comparison with current electronic technology might be useful. The Na^+,K^+-ATPase essentially functions as a miniature battery, converting chemical energy (in the form of ATP) into electrical energy (in the form of a transmembrane ion gradient). Whereas in the case of the gated, or switched, ion

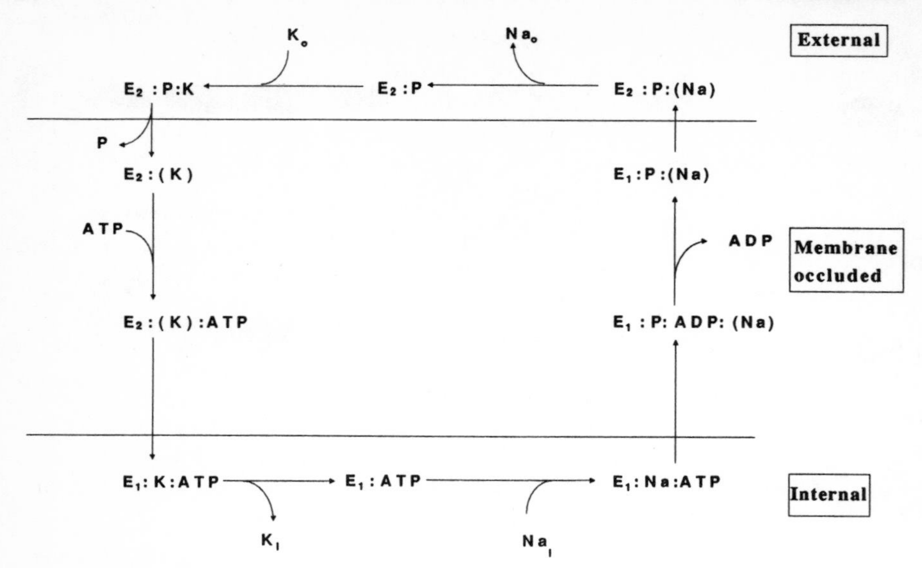

Fig. 2.6 Reaction scheme and model for the Na^+,K^+-ATPase. The 'o' and 'i' subscripts refer to the outside and inside of the cell, respectively. E_1 and E_2 refer to two states of the enzyme. P designates phosphate.

channels we were able to find a technological equivalent that was within three orders of magnitude of the biological device in terms of miniaturization, here the biological device has almost no equivalent within six orders of magnitude.

ENERGY CONVERSION IN BIOENERGETIC SYSTEMS

In our discussion of electron transporting chains of catalysts (above) we chose to put aside consideration of the energy-transducing roles of these chains and to concentrate instead on their properties as redox (electron transfer) proteins. Now we wish to turn our attention to the energetic aspects of these electron transfers.

Figure 2.3 shows not only a sequence of electron carriers but provides a relevant measure of how facile is the reaction between adjacent carriers. This measure is, of course, the standard reduction potential (or redox potential) measured with respect to the normal hydrogen electrode. The redox potential concept enables us to arrange the electron carriers of a naturally occurring chain into a sequence with the best reducing agent (best electron donor) as the primary donor and the best oxidizing agent (best electron acceptor) as the final acceptor.

The standard redox potential is, of course measured under thermodynamic standard conditions, the reduced and oxidized forms of the species in question being present in equimolar (1 M) concentrations. A 'mid-point' potential ($E_{m,7}$) is also usefully defined in which again equimolar concentrations of reduced and oxidized species are present, but the pH is 7.0 rather than the thermodynamically rigorous value (1 M concentration of hydrogen ions). For a system in which no protons are involved in the redox process (e.g. cytochromes) the mid-point potentional will have the same value as the standard redox potential. For a system in which protons are involved (e.g. quinone/quinol reactions), the effect of moving the pH through seven units away from standard conditions is to increase the redox potential by 7 × 60 mV, i.e. +420 mV, since the reduction will be less favoured. In a real system in which the concentrations or activities of the oxidized and reduced forms of the carriers can have non-standard values, the actual redox potential will differ from the standard or mid-point redox potential according to equation 2.1

$$E_{\mathrm{h}} = E_{\mathrm{m},7} + RT \ln\left(\frac{[\mathrm{ox}]}{[\mathrm{red}]}\right) \tag{2.1}$$

In practice, however, ordering carriers according to standard redox potentials gives a sequence which agrees well with experimentally determined functional sequences (see below for methods used to elucidate sequences kinetically). Figure 2.3 shows the values of standard and actual redox potentials for the mitochondrial electron-transport chain under 'energized' conditions (both NADH, the electron donor and oxygen, the acceptor, present in large excess). Although a number of components have shifted significantly in redox potential from their standard values (generally to more oxidized values) the sequence has not undergone any change.

The size of the redox potential step between adjacent carriers is a measure of the spontaneity of the reaction between the two components. Where this redox potential step is large, a significant amount of energy is available as a result of the electron transfer (the system loses energy since the electron falls from high to low potential). This energy may be simply dissipated as heat, but in the intact mitochondrial electron-transport chain, the energy available in the three largest steps (indicated in Fig. 2.3) is used to synthesize new chemical bonds in the formation of adenosine triphosphate (ATP).

It is of interest to compare the energy available from electron transport and that required for ATP synthesis. Under steady-state conditions prevailing in mitochondria, with a relatively high ATP concentration (5–10 mM) the energy required to synthesize ATP is of the order of 60 kJ/mol. The redox potential difference between the electron donor (NADH, $E_0 = -0.8$ V) and the electron acceptor (oxygen, $E_0 = +0.6$ V) is 1.4 V. Expressed as free energy available per mole of electrons passing down the chain this represents 135.1 kJ. This is usually expressed as energy per two electrons, since the donor, NADH is a two-electron donor. Thus we have 270 kJ available per two electrons transferred down the chain. If the system were 100% efficient, we would expect to see the synthesis of 270/60 = 4.5 moles of ATP per mole of NADH oxidized. In fact the measured rate of ATP synthesis is 3 moles per mole NADH, suggesting a thermodynamic efficiency of 66%, with some 33% of the energy available from the redox process dissipated in the form of heat. Clearly an efficient coupling mechanism is present between electron transport and ATP synthesis, and we will be concerned to discover the nature of this coupling mechanism below.

METHODS OF STUDY OF BIOENERGETIC SYSTEMS

Visible spectroscopy
Study of the function of bioenergetic systems is enormously helped by the fact that a number of the key components in energy-transducing electron transport chains have well-defined absorption peaks in the visible part of the spectrum. It is thus easy to literally see the process of oxidation and reduction in a molecule such as cytochrome *c*, whose absorption spectrum in the oxidized and reduced forms is shown in Fig. 2.7. The clear colour change which takes place on transition from one redox state to another is useful in at least two types of study. It can be used to identify potential electron donors and acceptors to the chain (or to a segment of the chain). Also, in conjunction with specific inhibitors of electron transport, the redox changes in a specific component in the presence and absence of the inhibitor can be used to either place components in sequence (if the point of action of the inhibitor is known) or (if information on the sequence is available) the point of action of the inhibitor can be identified. With rapid scanning spectrophotometers it is possible to measure changes in the redox state of more than one component simultaneously, although at room temperature the peaks of various components tend to merge making interpretation of these spectra difficult. Low-temperature spectroscopy (at liquid nitrogen temperatures) resolves peaks well, but since samples are frozen, it is not easy to make additions to change redox states of the components in real time. An interesting exception to this occurs with the light-activated systems, which can be activated in the solid state by a flash of light, following which, the changes in redox state of various components can be spectroscopically studied.

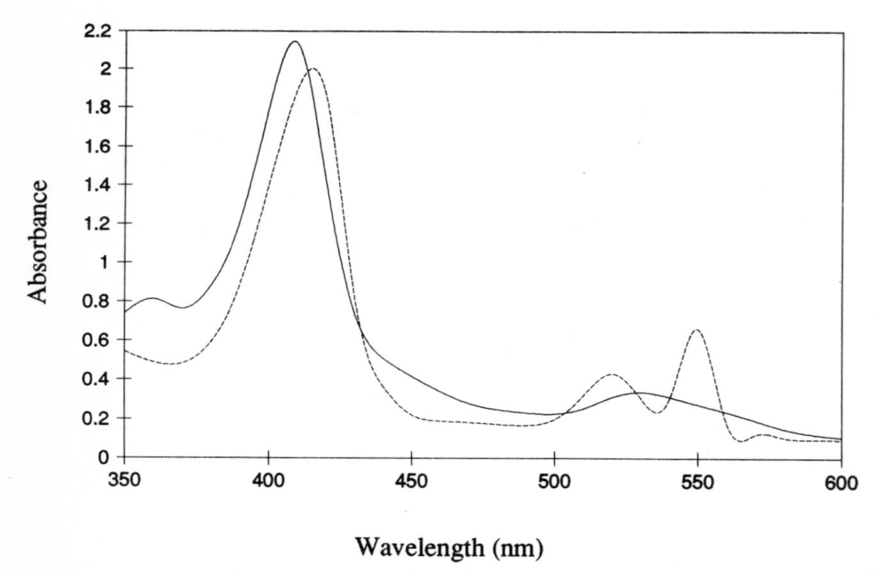

Fig. 2.7 Visible spectra of cytochrome *c* in the oxidized (——) and reduced (---) forms.

The aim of studies which use these techniques now is to identify the components which are involved in electron transport at the earliest possible stages following the primary, light-activated excitation of the photosynthetic reaction centre. The state of the art for these very fast response spectrophotometers are machines capable of resolution in time of the order of a few picoseconds.

Electron-spin spectroscopy

Although a number of interesting components of electron-transfer chains do have useful diagnostic visible absorption spectra, many other components which undergo redox reactions have no signals in this region of the electromagnetic spectrum. The redox state of many of these molecules can, however, be probed with a technique known as electron spin resonance (e.s.r.) spectroscopy, also known as electron paramagnetic resonance (e.p.r.) spectroscopy. For this technique to be applicable, the redox change should lead to a transition from a state in which the molecule possess unpaired spins (which exhibit the resonance phenomenon) to one in which the molecule is spin-paired or vice versa. Components showing useful e.s.r. signals include the iron–sulphur proteins, and the various quinone components.

Redox potentiometry

The various spectroscopic techniques alluded to in the previous sections provide a measure of the extent to which the various electron-transferring components are in their reduced or oxidized state. However, using these techniques in combination with various electron donors and acceptors and inhibitors provides only indirect evidence of the redox potential values. The technique of redox potentiometry, however, often used in conjunction with the spectroscopic techniques, provides a more direct means of establishing the redox potential of the components. The technique is an electrochemical one, in which a metal electrode (usually platinum) at known potential with respect to a reference electrode is used to donate/accept electrons to/from the species of interest. If used in conjunction with spectroscopic measurements, the potential at which the concentrations of the reduced and oxidized species are equal can be measured, and this is equal to the mid-point potential (see above). The instrumentation required for such measurements is a potentiostat and the principle of its operation is described below.

The redox potentiometry technique is best applied to purified components of the electron-transferring chains, but even where such components are available, direct redox potentiometry is often not possible, because of the very slow rate constants which characterize the reactions between large electron-transferring proteins and metal electrodes. These problems arise both because of the tendency of proteins to adsorb and denature at metallic surfaces and also because of the need for such molecules to orient precisely with the redox group facing the metal surface. For a large molecule such as a protein, the likelihood of an approach to the electrode surface at the precisely correct orientation is obviously less than for a small one. Small-molecule mediators are typically used to overcome this problem by acting as shuttle routes between the metal electrode and the protein in solution. This problem and current methods for attempting to overcome it is discussed further in Chapter 6.

Estimation of transmembrane electrochemical potential differences

A crucial part in many biological energy interconversions is played by various ion electrochemical gradients which are formed across membranes. Energy from chemical reactions such as ATP hydrolysis can be used to drive ion transport. Conversely the energy stored in ion electrochemical gradients can be used to drive thermodynamically unfavourable chemical reactions or other ion-transport processes. Direct interconversion between energy stored as an ion gradient and mechanical movement (the molecular 'motor') is also documented, as in the mechanism which drives the rotation of bacterial flagellae.

The electrochemical gradient which is most discussed and debated in bioenergetics is the proton electrochemical gradient which exists across the inner mitochondrial membrane (and other related membranes). According to the chemiosmotic hypothesis, this gradient is the obligatory intermediate between electron transport and ATP synthesis. That is, according to this hypothesis, the electron-transport chain uses energy from its exergonic redox reactions to pump protons across the membrane building up the proton electrochemical gradient, while the ATP synthetase (which is also located in the membrane) utilizes the energy stored in this gradient to accomplish the endergonic synthesis of ATP.

It is therefore important to be able to determine the magnitude of the electrochemical potential difference across the relevant membranes in any study of energy interconversions in bioenergetic systems. This involves the determination of two quantities, the difference in electrical potential ($\Delta\Psi$) and the difference in proton concentration (ΔpH) across the membrane. The measurement of both of these terms presents some difficulty in view of the fact that we are in general dealing with microscopic compartments which it is difficult to penetrate with even the smallest of microelectrodes, and in which chemical analysis is also difficult. In most cases, we resort therefore to indirect methods, which essentially rely on the measurement of the distribution of an indicator species between the two compartments concerned. As an example, we will look at the methods used to measure the electrochemical potential difference of protons across the mitochondrial membrane. This quantity is known as the protonmotive force, or p.m.f.

The p.m.f. across the mitochondrial membrane is generated by the action of the proton-translocating electron-transport chain. The mechanism by which this chain acts to translocate protons will be discussed below; for the present we need simply to acknowledge that it does so and that the direction of translocation is from inside the mitochondrion and into the cytoplasm. In the absence of other ion movements, therefore, the membrane potential will be negative inside. We therefore choose a positively-charged ion which will accumulate inside the mitochondrion to act as an indicator of membrane potential. A typical choice is $^{86}Rb^+$ in the presence of the ionophore valinomycin. The ionophore acts to allow the $^{86}Rb^+$ ion to equilibrate across the membrane. The distribution of the cation in the presence of the ionophore reflects the potential across the membrane, provided, of course, that the potential can be maintained by the proton-pumping system in the presence of the permeant cation. To ensure this condition is met, the concentration of indicator ion and ionophore is kept to a minimum. Since the ion is radioactive, it is a relatively simple matter to measure this distribution, following the separation of mitochondria from the suspending medium. The relationship between equilibrium ion distribution and membrane potential is given by the usual Nernstian equation:

$$\Delta\Psi = \frac{RT}{F}\ln\left(\frac{[\mathrm{Rb}^+]_\mathrm{i}}{[\mathrm{Rb}^+]_\mathrm{o}}\right) \tag{2.2}$$

According to this equation, if the ion is accumulated internally then $\Delta\Psi$ will be positive. A range of values for the ion distribution ratio and corresponding values of the membrane potential (mV) (in parentheses) is 2.00 (17.8); 4.00 (35.6); 8.00 (53.4); 12.00 (63.8); 16.00 (71.2); 20.00 (76.9).

Typically, in energized mitochondria (i.e. mitochondria which are well supplied with substrates which feed electrons into the electron-transport chain) this type of experiment yields values of about 60–100 mV for the membrane potential.

We now turn to the problem of determining the proton concentration ratio across the membrane. This question can also be resolved by use of an accumulating indicator species, but in this case we must measure the accumulation of a permeant weak acid (if the pH is lower inside) or a permeant weak base (if the pH is higher inside). In this case the appropriate relationship is:

$$\frac{[\mathrm{H}^+]_\mathrm{o}}{[\mathrm{H}^+]_\mathrm{i}} = \frac{[\mathrm{A}^-]_\mathrm{i}}{[\mathrm{A}^-]_\mathrm{o}} \tag{2.3}$$

where A^- signifies the anion formed in the dissociation of the weak acid. This relationship is derived from a consideration of the equal concentrations of the undissociated acid [HA] on either side of the membrane at equilibrium. Although it is not possible with tracer techniques to measure the ratio $[\mathrm{A}^-]_\mathrm{i}/[\mathrm{A}^-]_\mathrm{o}$, it can readily be shown that the ratio which *can* be measured $(([\mathrm{A}^-]_\mathrm{i} + [\mathrm{HA}]_\mathrm{i})/([\mathrm{A}^-]_\mathrm{o} + [\mathrm{HA}]_\mathrm{o}))$ (i.e. the ratio of the total concentration of ionized plus un-ionized forms of A) will approximate to the desired ratio at physiological pH values. Once both the concentration ratio and the membrane potential are known, the electrochemical potential can be calculated according to the relationship:

$$\Delta\tilde{\mu}_{\mathrm{H}^+} = \Delta\Psi + \frac{RT}{F}\ln\left(\frac{[\mathrm{A}^-]_\mathrm{i}}{[\mathrm{A}^-]_\mathrm{o}}\right) \tag{2.4}$$

This procedure leads to a value for the electrochemical potential difference for protons expressed in electrical units. The total p.m.f. estimated in this way in energized mitochrondria has been variously estimated to lie in the range 120–160 mV. How this transmembrane electrochemical potential can be used to drive other energy-requiring processes, and indeed whether this measure of stored energy is appropriate to all energy-transducing processes is discussed in the next section.

CURRENT THEORIES OF MECHANISMS IN BIOENERGETICS

Despite the fact that the methods described above can yield a good deal of information on parameters relevant to the functioning of energy-transducing membranes, we are still some way from a thorough understanding or a full description, in physico-chemical terms, of the mechanism by which the enzymes present in these membranes interconvert energy stores as described above.

Where the interconversion is relatively simple, as in the case of an ion-transporting ATPase (which essentially converts the chemical bond energy stored in the ATP molecule into the energy stored as an electrochemical gradient) there is at least some agreement on the direction in which research should proceed. Classical enzymological investigations, leading to the identification of intermediates in the catalytic cycle have already yielded insights into how such enzymes work. Thus for the plasmalemma Na^+,K^+-ATPase, a reaction sequence has been determined (in terms of ion binding, ATP hydrolysis, ion release), a series of well-defined intermediates have been found (including phosphorylated intermediates and intermediates in which ions are bound to enzyme sites in an occluded form), and key steps at which conformational changes in protein structure have been identified, although the precise nature of these conformational changes cannot yet be defined. The conformational changes in protein structure are of the greatest importance in understanding the functioning of these systems, although steps involving conformational changes are difficult to study and specify with precision, since this demands a knowledge of the three-dimensional structure of relevant parts of the protein molecule (if not the whole protein molecule) before and after the conformational change. In the case of many of the ion-transporting enzymes, it is the postulated conformational change which is responsible for exposing the ion to the opposite surface of the membrane. We shall return to this point when dealing with proton translocation later.

The situation is more complicated in those membranes where more complex energy interconversions are taking place. The mitochondrial membrane, for example, links redox energy, ion gradients and ATP. Thylakoid membranes add light energy to this list. In this area a basic controversy still exists: is the intermediate at the heart of all these energy interconversions a transmembrane proton electrochemical gradient, as required by the chemiosmotic hypothesis? While a glance at current biochemical standard texts would suggest that this was an accepted point of view, a scan through the relevant research literature shows major disagreements on this, and related matters.

Objections to the central dogma of chemiosmosis are essentially of two sorts. One group of critics suggest that a proton electrochemical gradient is an essential intermediate, but that we should not think of it as a gradient between the bulk phases on either side of the membrane – instead it should be thought of as a gradient which exists in a very localized environment close to the two membrane surfaces, on the surfaces, or even as charge separation within the membrane. In this view, the bulk-to-bulk proton electrochemical gradient, which is the only parameter that we can measure through measurements of $\Delta\Psi$ and ΔpH, is best thought of as the result of a 'leakage' of protons away from the membrane-localized environment.

A more radical criticism comes from another group of authors, who maintain that neither a bulk nor a localized proton electrochemical gradient is an *essential* intermediate in energy transduction. Instead, our attention should be focused on

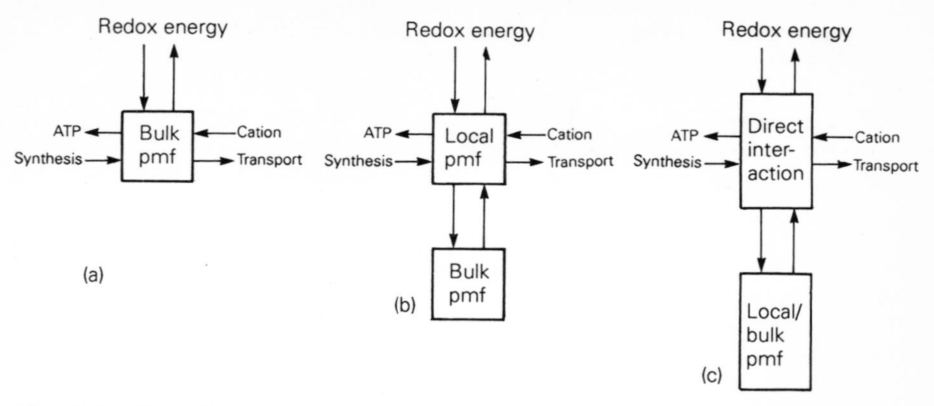

Fig. 2.8 Hypotheses for energy transduction in bioenergetic membranes. p.m.f. = protonmotive force. (a) Classical chemiosmosis; (b) Localized chemiosmosis; (c) Direct interaction.

direct interactions between (in the case of the mitochondrial membrane) the redox enzymes, responsible for exergonic electron transport, and the ATP synthetase, responsible for endergonic ATP synthesis. The proton electrochemical gradients, which are indisputably measurable, are seen by these radical critics as a non-essential 'backup' energy store. The three views – 'classical' chemiosmosis, 'localised' chemiosmosis and direct interaction – are illustrated in Fig. 2.8.

If we are interested in the exploitation of bioelectronic systems in devices, then the directly-interacting or highly-localized systems may look more attractive as systems offering potentially useful materials, since it reminds us more closely of our own solid-state devices than the other models which interact only via an aqueous phase. Which of these models is actually correct is, of course, yet to be decided, but there does seem to have been a discernible shift in opinion away from 'classical' chemiosmosis in recent years.

Bioelectrochemistry – the interface with technology

Electrochemical methods provide the means by which electronic devices and circuitry can communicate directly (through the passage of d.c. currents) with electroactive biomolecules. If bioelectronic devices are to be constructed which exploit the properties of biological molecules, it is likely that in many cases electrochemical methods and concepts will be involved at the interface between the electronic and biological components.

ELECTROCHEMICAL THEORY

Most of the redox reactions which we have considered thus far have taken place between reactants which are either in aqueous solution, or in the lipid environment of the membrane. In either case, we cannot alter the redox potential of the reactants except by manipulation of the oxidized/reduced ratio. Electrochemistry deals with those redox reactions in which the electron(s) is donated by, or is passed onto, a metal electrode immersed in the solution, so that only one of the redox reactants is a species in solution. One important practical aspect of electrochemical systems is that we can readily change the potential of the electrode with respect to a reference electrode using external circuitry. This enables us to obtain information about the redox properties of the species in solution by varying the electrode potential and studying the flow of current to/from the solution species.

The species in solution is able to undergo a transition from reduced to oxidized states and vice versa:

$$A_{ox} + e^- \rightarrow A_{red}$$
$$A_{red} \rightarrow A_{ox} + e^-$$

The starting point for electrochemical theory is the Nernst equation:

$$E = E_o + \frac{RT}{nF} \ln\left(\frac{[A_{ox}]}{[A_{red}]}\right) \quad (2.5)$$

which provides a relationship between E, the electrode potential, E_o, the standard oxidation–reduction (redox) potential of the solution species and $[A_{ox}]/[A_{red}]$ the ratio of the concentrations of the oxidized and reduced forms of the species in solution. If we are controlling the potential of the electrode by means of external circuitry then we can control the $[A_{ox}]/[A_{red}]$ ratio. Conversely, if we simply measure E, the electrode potential, without forcing it to any particular value, it will reflect the current value of the $[A_{ox}]/[A_{red}]$ ratio in solution at the electrode surface.

This view of electrode potentials is essentially a naive thermodynamic view, inasmuch as it assumes that the electrode and the solution have reached equilibrium and no net current is flowing. In practice, of course it is not always the case that an imposed change in electrode potential is immediately reflected in a change in the concentration of reduced and oxidized species in solution. That is to say, it takes time to reach electrochemical equilibrium. Where equilibrium is rapid, we describe such a system as *reversible* or *Nernstian.* Where the kinetics of electron transfer do not favour rapid equilibration, the system is referred to as *non-reversible* or *non-Nernstian.* It is clear that whether a system is regarded as Nernstian or non-Nernstian depends not only on the system itself, but also on the observer's resolution in time. A system which is Nernstian in the millisecond time scale may be non-Nernstian at the nanosecond scale. The relative speed with which the various systems reach equilibrium depend on the rate constants for the various electron-transfer reactions. Just as in homogeneous solution reactions, potential barriers have to be overcome before a reaction can take place, so for the heterogeneous electron-transfer reactions there must be an approach to the electrode of the electroactive species, and an appropriate orientation before an electron transfer can take place. Electrode kinetics are a major area of investigation in electrochemistry.

As we shall see later, these factors can become very important for large biological molecules, to the extent that their reactivity at metal surfaces can be very low indeed. For the present we shall simply note that two extreme forms of behaviour can be envisaged at any given electrode surface. In the first case we can imagine an electrode at which few electroactive species are present and for those that are present, kinetic features are so limiting that effectively no current will pass over a reasonably large range of potentials. An electrode behaving in this way is referred to as an ideal polarized electrode. It is characterized by a flat current vs. potential curve over a relevant range of potential. The opposite type of behaviour is exhibited by the ideal non-polarizable electrode, in which even small changes in potential give rise to current flows. Here the electrode reactions are extremely facile, and (if we limit the current to low values and provide large pools of electroactive materials) we will have an electrode whose potential does not change significantly with the passage of current. Such an electrode is practically realizable in the saturated calomel electrode design, which is for this reason frequently used as a reference electrode.

The historic terms 'polarizable' and 'non-polarizable' are in fact not the most appropriate for the phenomena being described, since in fact all electrodes show a polarization phenomenon, which we will now describe. This electrode phenomenon, which is quite independent of the electron-transfer reactions which can take place at the electrode surface, becomes apparent whenever a metal electrode at some definite potential is placed in an ionic environment. The polarization arises because the mobile ions in the aqueous phase are attracted to the oppositely charged electrode surface. This results in the formation of an ionic 'double layer' in which the excess electronic (or 'electron hole') charge in the solid phase is balanced by the ionic charges in the aqueous phase. In general, the surface charge density on the electrode is higher than can be generated by the counter-charges in the aqueous phase. We can therefore recognize two components which make up the double layer in the aqueous phase: there are the ions close to the electrode surface (at the inner Helmholtz plane) which make up the so-called 'compact layer'. Included among the species in this layer are solvent molecules and specifically adsorbed ions. In the outer, or 'diffuse' layer are those ions which are held by long-range electrostatic forces (at the outer Helmholtz plane and beyond). This double layer structure is shown in Fig. 2.9.

The existence of the ionic double layer leads to complications when measuring current flow in electrochemical experiments. Essentially two processes can be responsible for current flow: (i) charge transfer *across* the electrode surface and (ii) charge accumulation ('charging' of the double layer capacitance) *at* the electrode surface. The double-layer capacitance can have a large value in solutions containing high concentrations of electrolyte and it can be difficult to distinguish between the charge transfer current (the 'Faradaic' current), which is usually of more interest to the electrochemist, and the double-layer charging ('non-Faradaic') current, which is usually seen as an artefact or interference with the measurement being made.

The fundamental measurement in an electrochemical experiment is that of electrode current. We are most often interested in the way that the current varies

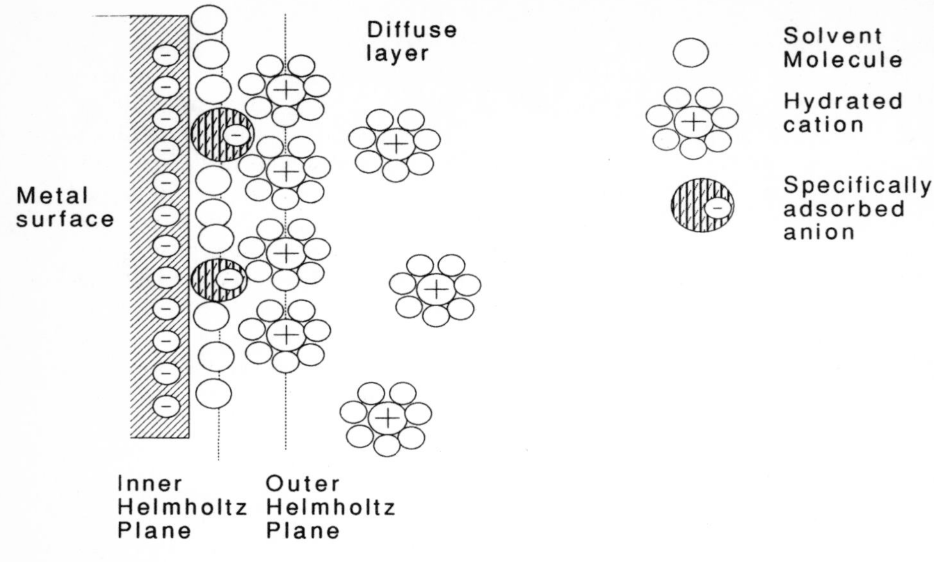

Fig. 2.9 Structure of the ionic double layer.

with applied potential and time. In general there are three factors which will affect the rate at which a Faradaic reaction can take place: (i) the rate of supply of electroactive species to the electrode surface (mass transfer); (ii) the rate of the heterogeneous electron-transfer reaction itself (electrode kinetics); and (iii) the progress of any obligatory-linked chemical reactions and physical processes (such as adsorption and desorption) which precede or follow the electrode reaction. We will consider these three factors in turn.

Mass transfer

A major difference between electrochemical reactions and 'normal' liquid-phase reactions is that the former take place only at a fixed surface, whereas the latter may take place anywhere in the bulk phase where the appropriate molecular collisions occur. It therefore follows that if the electrode reaction itself proceeds rapidly, the liquid phase immediately adjacent to the electrode surface will become depleted with respect to the electroactive species and an inhomogeneous distribution of reactants develops unless the solution is stirred vigorously. A reaction which initially proceeds rapidly at the electrode surface (following, for example, a step in potential) soon slows down as the concentration of electroactive species at the surface falls. The reaction is then said to be mass-transport controlled or limited, which for an unstirred solution and a neutral electroactive species, means *diffusion* controlled. Clearly when the solution is being stirred, convection may be the limiting factor and, where a charged electroactive species is involved, migration in the electric field will be an added factor.

When the concentration at the electrode surface of an electroactive species is depleted, a diffusion layer is formed between the bulk solution where the concentration has a constant value (C), and the electrode surface where the concentration can essentially be considered to be zero. The thickness of this diffusion layer can be shown to increase in proportion to $t^{1/2}$, where t is the time from the start of the reaction (the potential step). A characteristic of diffusion-controlled reactions is therefore that the current at constant potential declines with time as $t^{-1/2}$, since the diffusion layer expands at this rate. The relationship is expressed in the Cottrell equation:

$$i(t) = \frac{nFAD^{\frac{1}{2}}C}{\pi^{1/2}/t^{1/2}} \tag{2.6}$$

It is useful to bear in mind the actual dimensions of the various layers which we have described in the vicinity of the electrode. The diffusion layer is obviously a variable dependent on the time since depletion began, but for the times usually encountered in electrochemical experiments the range will be 0.5–50 μm. The ionic double layer, by contrast, will be 30 nm (0.03 μm).

If the solution is stirred, of course, then the concentration of electroactive species is maintained at a steady value at the electrode surface and the $t^{-1/2}$ dependency is not seen. Instead, the current will remain constant for a long period of time (until the bulk concentration becomes significantly depleted). The current–potential relationship for a Nernstian reaction is given by

$$E = E_{1/2} + \frac{RT}{nF} \ln\left(\frac{(i_l - i)}{i}\right) \tag{2.7}$$

where i_l is the limiting current and $E_{1/2}$ is the potential at which the current reaches half of its limiting value (see Fig. 2.10).

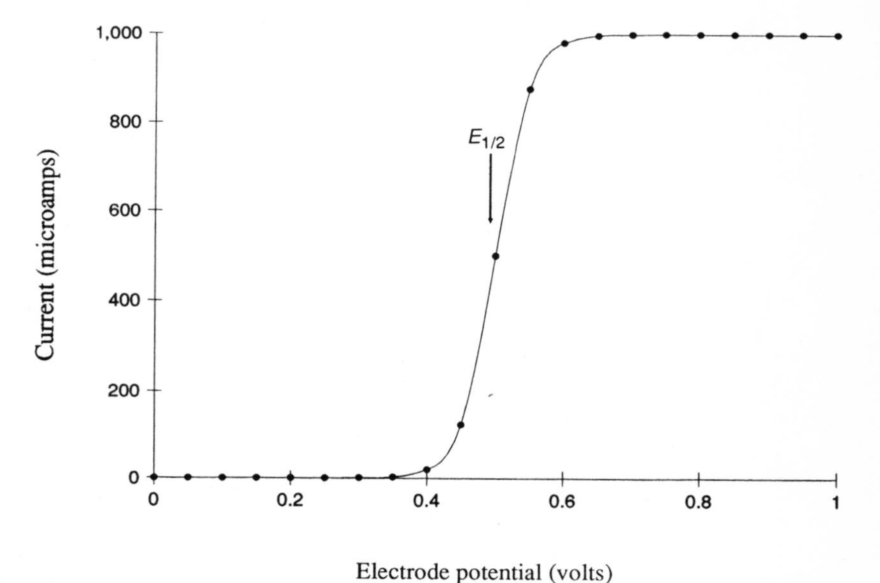

Fig. 2.10 Current–potential relationship for a stirred reaction. The limiting current in this case was 1 mA and the $E_{1/2}$ value was 0.5 V.

Electrode kinetics

The above discussion assumes that the rate of the electron transfer at the electrode is not limiting, such that transfer of molecules to the surface limits the rate of reaction. This is not always the case, however. If the kinetics of the electrode reaction are controlling the overall rate of reaction (as, for example in a stirred cell) then the current will not decay with time (as described above) and will be constant with time, but will be a sensitive function of what is known as the over-potential, η. The over-potential is defined as the difference between the actual potential, E, and the equilibrium potential E_{eq} which should be observed given the concentrations of reduced and oxidized species, so that $\eta = E - E_{eq}$. It has long been an established experimental fact that the current, i, in such a situation is related exponentially to the over-potential, thus:

$$i = \exp\frac{\eta - a}{b} \tag{2.8}$$

where a and b are empirical constants. This equation is often referred to as the Tafel equation and was originally formulated by Tafel in 1905 in the logarithmic form:

$$\eta = a + b \ln i \tag{2.9}$$

This should be contrasted with the case of the rapid electron-transfer reaction discussed above, since there is no limiting current value here. It can be shown that an equation of this form can be derived from first principles by considering electrode processes similarly to homogeneous solution reactions, with the additional consideration of the effect of altering the electrode potential on the activation energy of the reaction. For a well-stirred solution in which mass-transfer effects can be neglected, the theoretical treatment shows that the current is related to the over-potential as follows

$$i = i_0[e^{-\alpha nf\eta} - e^{(1-\alpha)nf\eta}] \tag{2.10}$$

known as the Butler–Volmer relation. Here i_0 is known as the exchange current, $f = F/RT$, and α is the transfer coefficient, a measure (value between 0 and 1) of the symmetry of the energy barrier which must be surpassed for a reaction to proceed, that is to say a measure of the relative effect which a given over-potential will have on the reductive and oxidative processes, respectively.

The relationship can be used to determine the parameters α and i_0 from experimental values of i and η. A Tafel plot, in which the natural logarithm of the observed current is plotted against the over-potential will show two straight line regions of slope $(1-\alpha)nf$ and αnf (see Fig. 2.11). The linear regions extrapolate to a value of $\ln(i_0)$ as the intercept on the $\ln(i)$ axis. Thus α can be determined from the slopes and i_0 from the intercepts.

Coupled chemical reactions

Where chemical reactions are coupled to the electron-transfer reactions (e.g. complexation of reduced ions), we can distinguish two cases – one in which the chemical reaction is itself rapidly equilibrating and one where the coupled

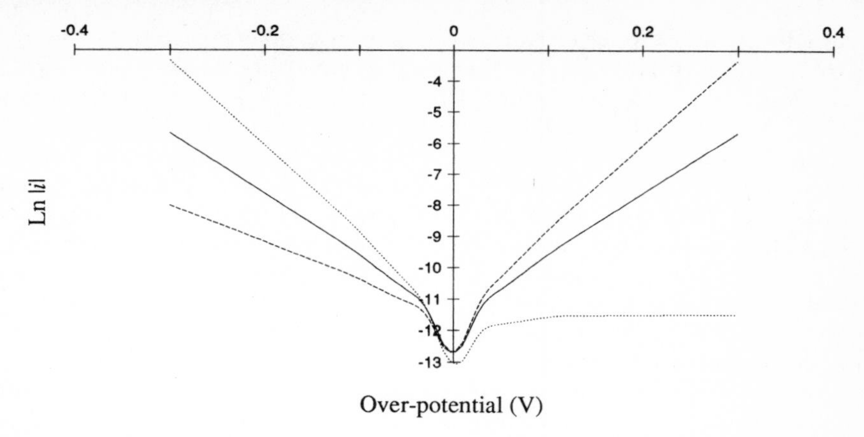

Fig. 2.11 Tafel plot for three different values of α: ——, $\alpha=0.5$; – – –, $\alpha=0.3$; . . . . ., $\alpha=0.7$. i_0 was in this case 10 μA.

reaction is slow. In the first case the *i–E* characteristic is still Nernstian, but the $E_{1/2}$ value is shifted by a value that depends on the equilibrium constant for the chemical reaction and on the concentration of the reacting species. In the latter case, the Nernstian *i–E* characteristic is still seen, but again is shifted by a factor which depends now on kinetic rather than thermodynamic properties of the coupled reaction.

ELECTROCHEMICAL METHODOLOGY

Electrochemical methods are often thought of as mainly useful in quantitative analysis. Such analytical techniques are not irrelevant to bioelectronics, being an important area in biosensor design. Here, however, we will give a brief survey of a series of electrochemical methods which can be used to study the nature of reactions occurring at electrode surfaces without any emphasis on detecting the concentrations of the species involved.

Although the fundamental measurement in electrochemistry is nearly always one of current (vs. time or potential) there are many ways of making such a measurement depending on (i) the properties of the system we are dealing with and (ii) the type of information we are attempting to obtain. Some of the various methods will be described below. Readers are referred to excellent texts for a full treatment.

The potentiostat

In all our discussions of electrochemical reactions we have concentrated on the process occurring at a single electrode (usually termed the 'working electrode'). Obviously, to make a current flow through a solution we require a minimum of two electrodes, the second usually referred to as the 'counter-electrode'. We are not usually interested in the electrochemical processes taking place at the counter-electrode, but we need to be sure that whatever they are they do not limit the current flow at the working electrode. This is usually achieved by arranging for the counter-electrode to have an area much greater than that of the working electrode. If we also need to know the potential of the working electrode, then we could conceivably use as the counter-electrode a reference electrode of fixed potential (e.g. a saturated calomel electrode (SCE)). However, even designs such as the SCE do not maintain fixed potentials over a wide range of current values. We therefore have a problem with such a two-electrode cell design in measuring current at a known potential. Another problem with such a cell arises due to the finite resistance of the solution between the electrodes R_s. When a current passes through the solution there will be a voltage drop equal to iR_s, which means that the potential of the working electrode will be less that the applied potential by this amount. Where i is a constant, this simply implies that our potential scale is shifted by iR_s, but where i varies with time, clearly there will be a variation in the error with which the working electrode potential is determined.

The answer to both of these problems (iR drop and drawing too high a current through the reference electrode) lies in the use of the three-electrode potentiostat. This instrument (illustrated schematically in Fig. 2.12) draws virtually no current through the reference electrode, but instead uses a third electrode as a counter-electrode. Thus the potential of the working electrode can be maintained at the desired level with respect to the reference, but no current is drawn through the reference electrode allowing it to maintain a constant potential. The circuit operates by adjusting the potential at the counter-electrode to whatever value is necessary to maintain the desired potential between the working and reference electrodes. At the same time, the current at the working electrode is measured. The

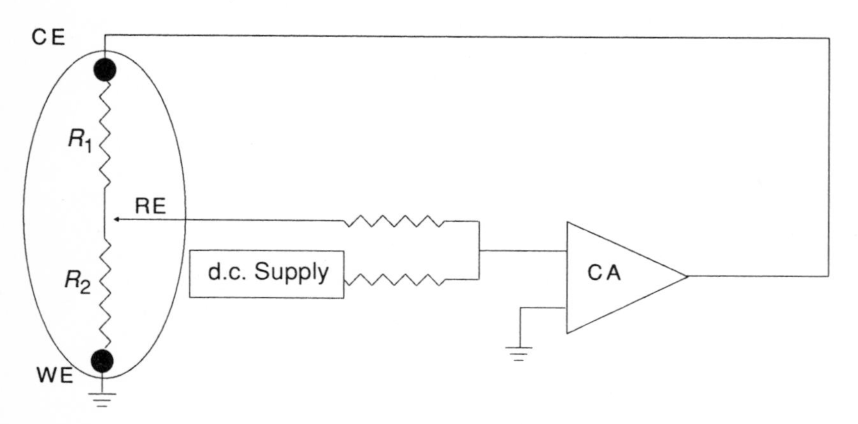

Fig. 2.12 Schematic diagram of a potentiostat. CA = control amplifier. CE, RE and WE refer to counter, reference and working electrodes respectively. R_1 and R_2 refer to solution resistances. The value of the potential supplied by the d.c. supply sets the potential of the WE with respect to the RE.

problem of solution resistance is solved to the extent that whereas in a two-electrode cell, the relevant resistance (see Fig. 2.12) would be $(R_1 + R_2)$, in the three-electrode design, only R_2 presents a problem. This can be further minimized by placement of the working electrode as close as possible to the reference electrode. In modern, microcomputer-controlled, potentiostat designs, further refinements are possible to compensate for this residual 'uncompensated' solution resistance.

With a working potentiostat we can begin to study electrode reactions with a whole battery of techniques. We will describe only the most common and most useful for our purposes. In general, in applying these techniques to biological problems, the questions being addressed are: (i) do the biomolecules in question undergo electron-transfer reactions at the electrode? (ii) are the reactions rapid (showing Nernstian behaviour) or are they slow? (iii) if we are dealing with non-Nernstian systems, what are the values of the kinetic constants?

Linear sweep and cyclic voltammetry

In the linear-sweep technique, the voltage applied to the working electrode is varied linearly (with time) from an initial value to some final value while current is monitored. For an unstirred solution, if we begin at a potential at which no electrochemical reaction takes place, a peak current will be seen at a particular potential (E_p), which is related to the redox potential for the electroactive species which is reacting. The reason for the fall in current at potentials above E_p is that diffusion limitation begins to play an important role in determining current. A half-peak potential can also usefully be defined ($E_{p/2}$) at which the current is half of its peak value. Anodic (oxidative) and cathodic (reductive) sweeps are obviously possible and with the correct choice of limiting values for the applied potential, oxidative and reductive peaks will be observed. Expressions have been derived which state where these two peaks should lie with respect to each other for a reversible (Nernstian) system. Linear-sweep voltammetry may therefore be used to diagnose a system as being either Nernstian or non-Nernstian.

Cyclic voltammetry is an experimental method in which cathodic and anodic linear sweeps are performed one immediately after another. Thus the (relatively small) amount of product which is formed in the vicinity of the electrode surface on the first sweep is removed on the reverse sweep. Obviously from a cyclic voltammogram it is immediately apparent if the product formed on the forward sweep is capable of removal on the reverse sweep. The lack of a peak in the reverse sweep may be due to the kinetic constants for the reverse reaction being unfavourable (large or small values of α) or due to the product being in some way highly reactive and being removed by chemical reaction before electrochemical reversal has taken place.

Cyclic voltammetry is perhaps the easiest diagnostic technique for distinguishing Nernstian from non-Nernstian behaviour. Nernstian reactions have a characteristic peak separation ΔE_p of $58/n$ mV (where n is the number of electrons involved). Where peak separation is much greater than this value, kinetic parameters (particularly k_0, the standard heterogeneous rate constant for the electron-transfer reaction) may be determined from the value of ΔE_p.

Hydrodynamic methods. The rotating disk electrode

In these methods diffusion plays a much diminished role in determining the current flow at the electrode since the solution is stirred. The rotating disk electrode is used to achieve a steady state current which can easily and accurately be determined. Since steady-state, rather than transient currents are measured, double-layer charging phenomena are much less of a problem. A diffusion layer is still present at the electrode surface, but is much less significant in determining current and is a function of the speed of rotation of the electrode.

An expression (the Levich equation) for the limiting current under mass-transfer limiting conditions in a rotating disk electrode experiment can be derived as follows:

$$i_l = 0.620nAD^{2/3}v^{-1/6}\omega^{1/2}C \qquad (2.11)$$

where n is the number of electrons involved in the reaction, A is the electrode area, D is the diffusion constant of the reacting species, v is the kinematic viscosity of the solution, ω is the angular frequency of rotation of the electrode and C is the bulk concentration of the electroactive species. Where kinetic limitations mean that the reaction is not mass-transfer limited (the current is not proportional to $\omega^{1/2}$) more complex expressions have been derived which contain k_f and k_b, the rate constants for the forward and backward electron-transfer reactions.

a.c. methods

Mention should finally be made of a group of methods in which a sinusoidally varying potential waveform is applied to an electrode. Here the current response will also be sinusoidal and there will be both a magnitude and a phase (with respect to the potential waveform) to measure. If the measurements are repeated across a wide range of frequencies, this procedure allows the complex impedance to be plotted as a function of frequency. A model circuit, consisting of capacitors and resistors can then be devised which has the same impedance vs. frequency characteristics. The model circuit components can then be related to physical parameters such as the double-layer capacitance and the charge-transfer resistance, again allowing the determination of kinetic constants.

APPLICATION OF ELECTROCHEMISTRY TO BIOMOLECULAR SYSTEMS

Among the many types of molecules which make up biological systems are a number which have electroactive properties. These have been surveyed by Dryhurst and some important specific electrochemical reactions of biomolecules are discussed in Chapter 6. We can divide the molecules of interest into small molecules and macromolecules, the latter referring primarily to proteins. While no special considerations apply to the low-molecular-weight compounds, the proteins are a special case because of their large size, and the localization of the electroactive group in a particular small portion of the large molecule. This implies that in general the reactions will be slow, i.e. kinetically-controlled rather than mass-transfer controlled. The techniques which will be applicable to these proteins will therefore be those which are capable of measuring kinetic parameters and comparing these parameters for various electrode surfaces and other experimentally manipulable conditions.

Another important consideration which applies to a large number of electroactive biological molecules is the fact that many of their electrode reactions are electron plus proton reactions rather than electron-only reactions. This implies that the electrochemical reactions will be pH-dependent. When comparing results, for example on related compounds reacting at various electrode surfaces, pH is a factor which must therefore be carefully controlled.

Electrochemistry therefore supplies techniques by which the electrochemistry of biological molecules can be studied and possibly exploited. Electrochemical theory can also however be useful in building new hypotheses to explain electrical and electrochemical phenomena in biological systems. Kell used electrode kinetic theory to substantiate the localized chemiosmotic hypothesis. Habib and Bockris have, for example recently presented an interesting new hypothesis on membrane potential generation in which the membrane behaves like an electrode, or rather different parts of the membrane behave like a cathode and an anode, transferring electrons to electroactive components in the medium. The validity or otherwise of this hypothesis is not important here – it stands simply as an example of the rich source of ideas which is provided by electrochemical theory and of the applicability of some of those ideas to biological electrical phenomena.

Selected reading

Bard, A.J. and Faulkner, L.R. (1980). *Electrochemical Methods.* New York, John Wiley & Sons.

Bockris, J.O'M. (1973). *Modern Electrochemistry.* New York, Plenum.

Dutton, P.L. (1978). *Meth. Enzymol.* **54**, 411–435.

Ferguson, S.J. (1985). *Biochim. Biophys. Acta* **811**, 47–95.

Finean, J.B., Coleman, R. and Michel, R.H. (1981). *Membranes and their Cellular Functions.* Oxford, Blackwell Scientific.

Habib, M.A. and Bockris, O'M. (1986). In *Modern Bioelectrochemistry*, Eds Gutman, F. and Keyzer, H. New York, Plenum, pp. 69–93.

Hille, B. (1989). *Ion Transport*, Eds Keeling, D. and Benham, C. London, Academic Press, pp. 57–69.

Kell, D.B. (1979). *Biochim. Biophys. Acta* **549**, 55.

Mathews, C.K. and Van Holde, K.E. (1990). *Biochemistry.* Redwood City, Calif., Benjamin/Cummings Inc.

McNab, R.M. and Aizawa, S.I. (1984). *Ann. Rev. Biophys. Bioeng.* **13**, 51–83.

Pethig, R. (1986). In *Modern Bioelectrochemistry*, Eds Gutman, F. and Keyzer, H. New York, Plenum, pp. 199–236.

Singer, S.J. and Nicholson, G.L. (1972). *Science* **175**, 720–731.

Chapter 3

Theory and Practice of Dielectric Spectroscopy

Introduction

Light absorption spectroscopy, the frequency-dependent absorption by materials of ultraviolet, visible and infra-red light, is nowadays a basic analytical tool in most laboratories. This technique uses electromagnetic energy with frequencies in the range 10^{14} to 10^{16} Hz to probe the atomic and electronic 'make-up' of molecules from their characteristic absorptions. In a similar way, electromagnetic energy of lower frequencies, typically 10^{-3} Hz to 10^{11} Hz, can be used to investigate rotational and charge-transport properties of materials from relaxation absorptions and this is known as *dielectric relaxation spectroscopy*.

Over the last forty years, dielectric relaxation spectroscopy, in its various forms, has developed into an important analytical tool for examining the structure and function of biological molecules. Dielectric measurements, made over a wide frequency range, can provide a variety of information about biomolecular systems. At the simplest level, for dipolar systems, parameters such as the effective dipole moment of relaxing dipoles and their characteristic relaxation times can be obtained. For heterogeneous systems where dielectric losses result from interfacial polarizations, dielectric spectroscopy can provide values for the conductivities and permittivities of the constituent phases. In the case where dielectric losses occur as a result of lateral charge migration, the number density and mobility of charge carriers such as protons and ions can be obtained. As we shall see in Chapter 5, dielectric spectroscopy can also be used to probe the structural flexibility of molecules such as proteins and membrane systems and is able to provide information about biologically relevant protein–ligand interactions.

In the first part of this chapter we will be looking at the types of polarization occurring in biological systems – orientational polarization of permanent and induced dipole moments and their rotational relaxation properties; Maxwell–Wagner interfacial polarizations which result from microscopic inhomogeneities; and polarizations due to counter-ion migrations. The second half of the chapter deals with the experimental side of dielectric spectroscopy, describing a number of techniques for acquiring dielectric data and discussing some of the problems which may be encountered.

Theory

ORIENTATIONAL POLARIZATION

Permanent dipoles in biological systems

A large number of biological molecules have, inherent in their structure, fixed charges separated by atomic distances which endow them with permanent dipole moments. The charges may be equivalent in magnitude to whole electronic charges as they are in the zwitterionic form of amino acids (Fig. 3.1) or to fractions of an electronic charge, produced as a result of induction effects, such as those possessed by water molecules (Fig. 3.2). In both cases the permanent dipole moment, μ, is given by

$$\mu = qd$$

that is, the product of the magnitude of the charge, q and the charge separation, d.

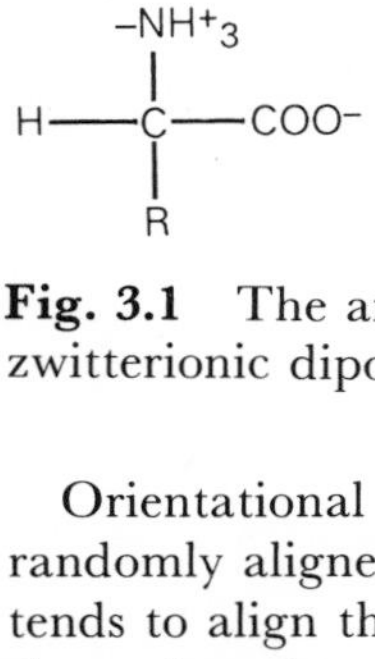

$-NH_3^+$
H—C—COO^-
R

Fig. 3.1 The amino acid zwitterionic dipole.

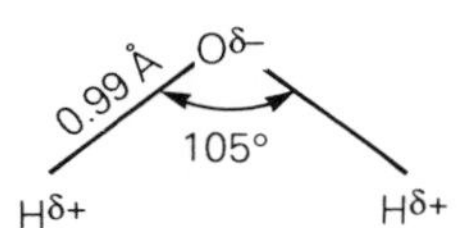

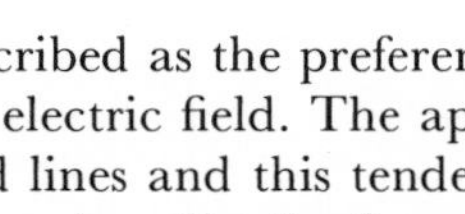

Fig. 3.2 The structure and charge distribution of the water molecule.

Orientational polarization can be described as the preferential orientation of randomly aligned dipoles in an applied electric field. The applied electric field tends to align the dipoles along the field lines and this tendency is opposed by thermal motions which tend to restore a random distribution of orientations. The relation linking the extent of the orientational polarization to the applied electric field and the temperature is given by the Langevin function $L(x)$,

$$\frac{\mu_d}{\mu} = \coth x - \frac{1}{x} = L(x) \tag{3.1}$$

where μ is the dipole moment of each molecule, μ_d is the average dipole moment per dipole in the direction of the applied electric field and the term x is given by

$$x = \frac{\mu E}{kT} \tag{3.2}$$

where E is the applied electric field, T is the temperature and k is the Boltzmann constant. A graphical representation of this function is shown in Fig. 3.3.

The term coth x in equation (3.1) can be written as

$$\coth x = 1/x + x/3 - x^3/45 + \ldots$$

For easily realizable electric fields (that is $E < 1 \times 10^7$ V/m) and typical values of dipole moment ($\mu \sim 1 \times 10^{-30}$ C/m) it can be seen that $x \ll 1$. Therefore terms involving x^3 and higher powers are negligible and the Langevin function $L(x)$ may be written as

$$L(x) = \frac{\mu_d}{\mu} = \frac{\mu E}{3kT} \qquad (3.3)$$

For realizable electric fields, the ratio μ_d/μ, which defines the Langevin function and which is a measure of the alignment of the dipoles in the electric field, is much less than unity. This implies that the polarization of the dipoles, that is the extent to which the dipoles are aligned, is only slight.

From equation (3.3) the average dipole moment μ_d may be written

$$\mu_d = \frac{\mu^2 E}{3kT} = \alpha E$$

where α is known as the polarizability.

Rotational relaxation

For those molecules possessing permanent dipole moments, the ability of the dipoles to orient in the applied electric field determines their dielectric properties. If an alternating electric field of sufficiently low frequency is applied to an

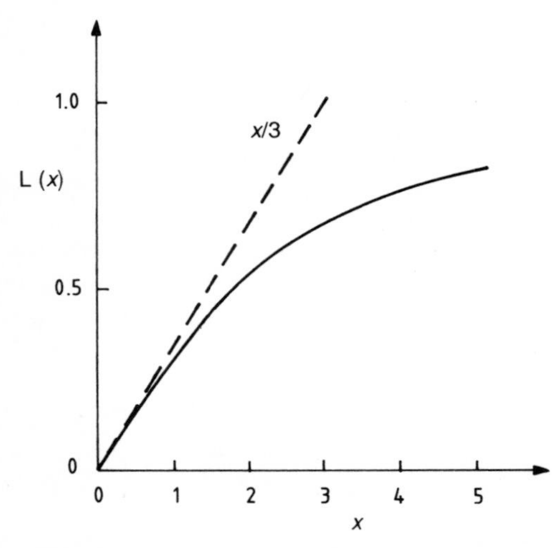

Fig. 3.3 Graphical representation of the Langevin function.

assembly of such molecules, then the dipoles are able to follow the changing electric field and the re-orientation of the dipoles transmits electromagnetic energy through the medium. The movement of the charges associated with each dipole produces a current in the dielectric which is termed the *displacement current*. At low frequencies when there is no lag between the orientation of the dipoles and the variation of the applied electric field, the displacement current is 90° out of phase with the voltage producing the electric field. At this stage there is no component of the displacement current in phase with the voltage and so the electromagnetic energy is transmitted without loss (that is, the product $V \times i = 0$). As the frequency is increased, a condition is eventually reached where the orientation of the dipoles can no longer completely keep up with the changing electric field. As a consequence of this, the displacement current acquires a component which is in-phase with the voltage producing the electric field and the product Vi becomes finite. As a result, Joule heating of the dielectric occurs and this phenomena is known as *dielectric loss*. In addition, since the dipoles are becoming unable to follow the changing electric field, there is a fall in the amount of charge stored by the dielectric as the frequency increases and this is reflected as a decrease in the permittivity. As the frequency is further increased over this region, the condition is eventually reached where the dipoles are completely unable to respond to the applied field. By this stage, the displacement current has fallen to zero resulting in a return to the zero loss condition and the charge storage capability becomes equivalent to that of a non-polar dielectric. The dielectric characteristics, in terms of permittivity, ε', and dielectric loss, ε'', for a typical dipolar molecule are shown in Fig. 3.4.

The Debye equation, which describe the relationship between permittivity, dielectric loss and frequency may be written in terms of the complex equation,

$$\varepsilon^*(\omega) = \varepsilon' - i\varepsilon'' = \varepsilon_\infty + \frac{(\varepsilon_s - \varepsilon_\infty)}{1 + i\omega\tau} \tag{3.4}$$

where $\varepsilon^*(\omega)$ is known as the *complex permittivity*, ε_s is the low-frequency relative permittivity, ε_∞ is the high-frequency relative permittivity, ω is the angular frequency and τ is the characteristic relaxation time given by

$$\tau = 1/2\pi f_r$$

where f_r is the frequency of the maximum dielectric loss.

By equating the real and imaginary parts, the permittivity and dielectric loss may be written as

$$\varepsilon'(\omega) = \varepsilon_\infty + \frac{(\varepsilon_s - \varepsilon_\infty)}{1 + \omega^2\tau^2} \tag{3.5a}$$

and

$$\varepsilon''(\omega) = \frac{(\varepsilon_s - \varepsilon_\infty)(\omega\tau)}{1 + \omega^2\tau^2} \tag{3.5b}$$

The above equations describe the rotational relaxation of a system exhibiting a single relaxation time with a dielectric dispersion with a half-height frequency

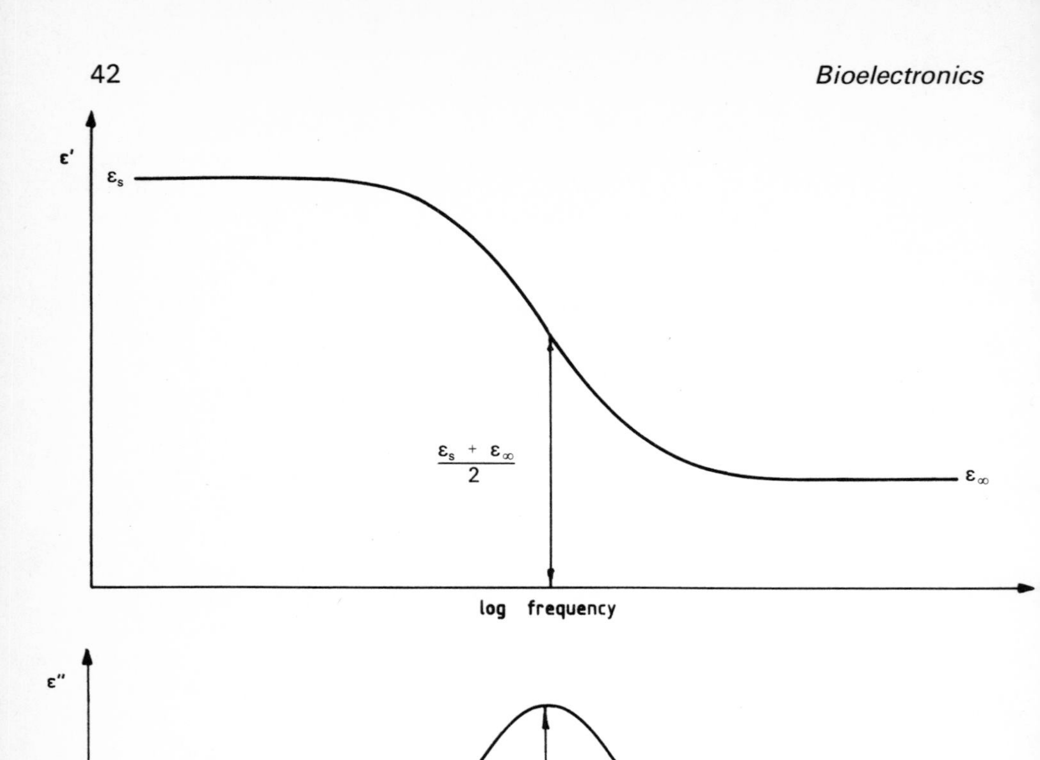

Fig. 3.4 The frequency dependence of the real and imaginary parts of the complex permittivity.

width of 1.14 decades. When a distribution of relaxation times exists, the width of the dispersion increases and the Debye equation is modified by the introduction of the parameter β as follows:

$$\varepsilon^*(\omega) = \varepsilon_\infty + \frac{(\varepsilon_s - \varepsilon_\infty)}{1 + (i\omega\tau)^\beta} \tag{3.6}$$

where $0 < \beta \leqslant 1$. For a system exhibiting a single relaxation time, $\beta = 1$ and β approaches zero for an infinite distribution of relaxation times.

In addition to the frequency plots shown in Fig. 3.4, dielectric data may be presented in terms of a so-called *Cole–Cole* (after the authors) or *complex permittivity*

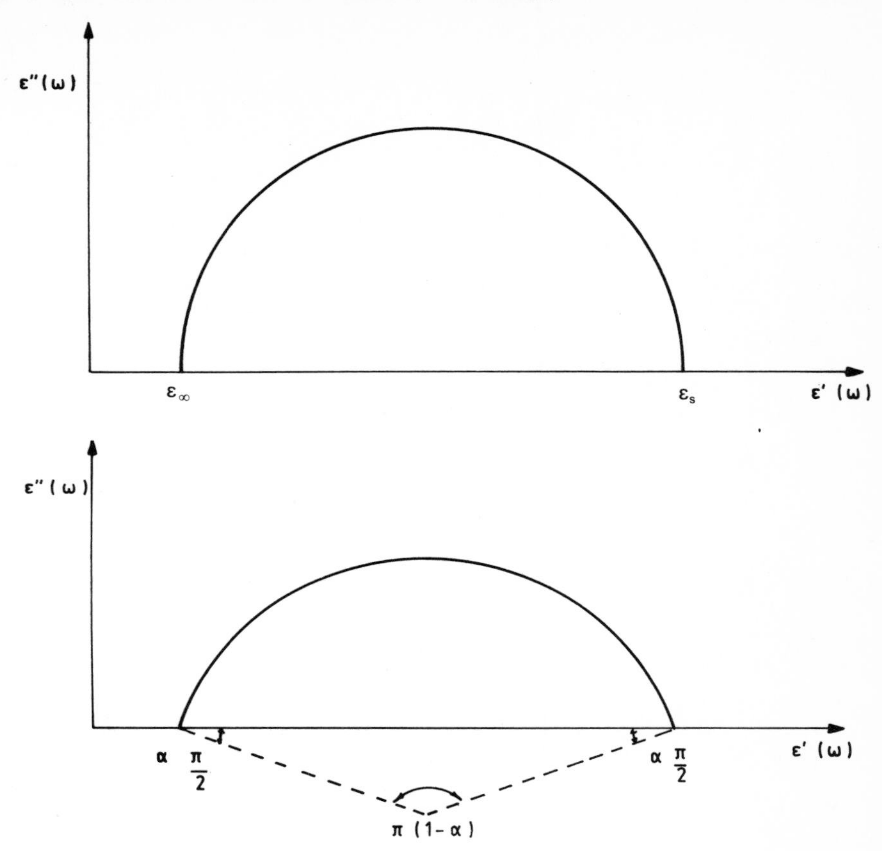

Fig. 3.5 Complex permittivity (Cole–Cole) plot.

plot in which ε'' is plotted against ε'. This type of plot usually takes the form of a semi-circle of the type shown in Fig. 3.5 with a centre which lies on, or below, the abscissa and which makes an angle $\alpha\pi/2$ radians with the points ε_s' and ε_∞'. It turns out that the parameter β from equation (3.6) is equal to $1-\alpha$.

For a relaxing system exhibiting a single relaxation time, the centre of the Cole–Cole plot lies on the abscissa and $\alpha=0$. The value of α becomes finite and the centre lies below the abscissa for systems exhibiting a distribution of relaxation times.

Davidson and Cole proposed a further modification to the Debye equation in order to characterize a non-uniform distribution of relaxation mechanisms. This relation is of the form

$$\varepsilon^*(\omega)=\varepsilon_\infty+\frac{(\varepsilon_s-\varepsilon_\infty)}{(1+i\omega\tau)^\gamma}$$

and has the form of a skewed arc, where $0<\gamma\leqslant 1$.

Debye used a simple physical model to provide a description of the rotational relaxation process. The dipoles are assumed to be spheres whose rotation in an electric field is opposed by the microscopic viscosity of the medium. The relaxation time, τ, for such a model is given by

$$\tau = \frac{3V\eta}{kT} = \frac{4\pi\eta r^3}{kT} \tag{3.7}$$

where V is the volume and r is the radius of the sphere and η is the viscosity of the medium. For ellipsoidal particles, different relaxation times will apply to rotations about any of the three axes. For the case of the prolate ellipsoid where $r_1 \gg r_2 = r_3$, the relaxation time characterizing rotation about the small axis is given by

$$\tau = \frac{8\pi\eta r_1^3}{3kT[2\ln(2p) - 1]}$$

where $p = r_1/r_2 \gg 1$.

In spite of the simplicity of the model, a high degree of agreement is evident between experimentally-obtained values of τ and the values calculated from the viscosity equation.

In addition to the relaxation time, an important parameter in dielectric spectroscopy is $\Delta\varepsilon'$, the dielectric increment which is defined simply as the difference between the low-frequency and high-frequency relative permittivities:

$$\Delta\varepsilon' = \varepsilon_s - \varepsilon_\infty$$

The experimentally-observed dielectric increment is related to the dipole moment of the polar molecules, μ, and the molecular weight M according to the equation

$$\Delta\varepsilon' = \frac{N\mu^2 g c_o}{2\varepsilon_o MkT} \tag{3.8}$$

where ε_o is a constant known as the permittivity of free space, N is Avogadro's number, and c_o is the concentration (kg m^{-3}) of the polar molecules. The parameter g, known as the Kirkwood correlation parameter, is introduced to account for local molecular interactions and correlation effects which occur, for example, between solute and solvent molecules. Such interactions can be very important in aqueous solutions of biological molecules where there may be extensive hydrogen bonding. An ionized amino acid may, for instance, hydrogen bond up to four water molecules whose rotational motions are correlated to that of the amino acid. Water itself is engaged in hydrogen-bond interactions with neighbouring water molecules and has a g value of 2.82 at 20 °C. For a system where there are no intermolecular interactions or where there are many diffuse interactions which effectively cancel (as they do for example for proteins in solution) then the value of g reduces to unity.

INDUCED POLARIZATIONS

This type of polarization arises as a result of local electric field-induced separation of charges which effectively results in the formation of new dipoles. Two important

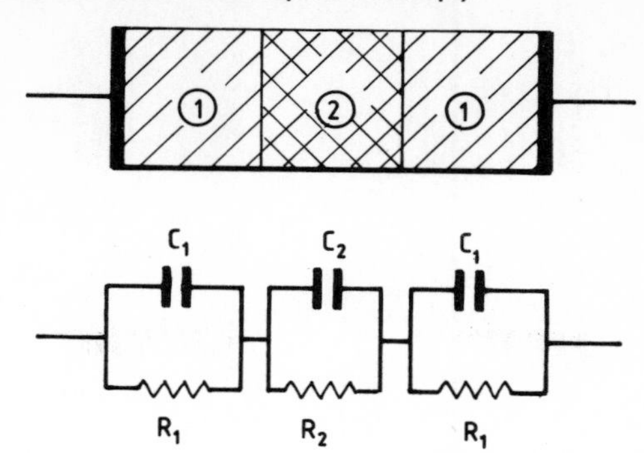

Fig. 3.6 (*Top*) Schematic diagram of a two-phase dielectric; (*bottom*) equivalent circuit representation.

examples of induced polarizations in biological systems are field-induced counter-ion migration on the surface of macromolecules and Maxwell–Wagner interfacial polarizations at non-conducting boundaries.

Maxwell–Wagner polarization

Biological materials are inherently microscopically heterogeneous, and are composed of molecules which exhibit considerably different permittivities and conductivities. When an electric field is applied to such a material, the mobility of charge carriers, such as ions, migrating through the material can be significantly higher in some regions (for example an aqueous phase) compared with other regions (for instance a lipid phase). This inevitably leads to a build up of charge carriers at non-conducting boundaries and results in a non-uniform charge distribution in this region. The heterogeneous system then exhibits frequency-dependent properties which are different from either of the constituent phases.

The theoretical aspects of the electrical behaviour of heterogeneous dielectrics have been investigated by Maxwell and Lord Rayleigh and later developed by Wagner. The simplest heterogeneous systems to be analysed consist of parallel layers of different dielectrics placed between two electrodes such as the three-layer configuration shown in Fig. 3.6. This system has been used as a model of the bimolecular lipid membrane with the two outer components being assigned to the conductive polar head group regions and the inner region corresponding to the hydrocarbon lipid tales, as described in Fig. 3.7. Using circuit analysis, the capacitance and conductance of the total system can be described by the following equations

$$C = \frac{C_H G_p^2 + 2C_p G_H^2 + \omega^2 (C_H C_p^2 + 2C_H^2 G_p)}{(G_p + 2G_H)^2 + (\omega C_p + 2\omega C_H)^2} \tag{3.9a}$$

$$G = \frac{G_H G_p^2 + 2G_H^2 G_p + \omega^2 (C_H C_p^2 + 2C_H^2 C_p)}{(G_p + 2G_H)^2 + (\omega C_p + 2\omega C_H)^2} \tag{3.9b}$$

where G_p and G_H, and C_p and C_H, are the conductances and capacitances of the polar head group and hydrocarbon phases, respectively.

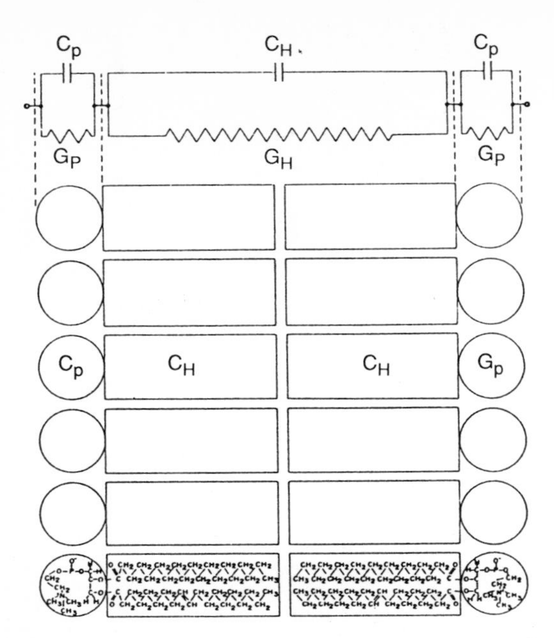

Fig. 3.7 Schematic diagram of a bimolecular lecithin membrane. The hydrocarbon (H) and polar head group (P) regions are assigned an equivalent capacitance and conductance producing the equivalent circuit shown at the top of the figure.
From Coster, H.G.L. and Smith, J.R. (1974). The molecular organisation of bimolecular lipid membranes, *Biochim. Biophys. Acta* **373**, 151–164.

Measurements of the capacitance and conductance of lecithin black lipid membranes have indicated a dielectric loss centred at approximately 5 Hz. By introducing values of conductance and capacitance for the head group and hydrocarbon regions into equations 3.9(a) and (b), good correlation between the Maxwell–Wagner model and the experimental data was obtained. Using this type of analysis, it is possible, for instance, to obtain information about the conductivity and permittivity of the membrane sub-structure.

Similar analysis can be applied to other heterogeneous systems, for example spherical insulating shells suspended in a conducting medium. This model is of particular relevance to cells in aqueous suspension and is discussed in Chapter 4.

Counter-ion relaxation

There are a number of instances in biology where charged groups exposed to the aqueous phase are compensated by counter-ions of the opposite charge. This phenomenon occurs, for example, in DNA where negatively-charged phosphate groups are directed outwards from the molecular backbone (Fig. 3.8a) and also in cell membranes composed from lipids with ionizable head groups (Fig. 3.8b). In

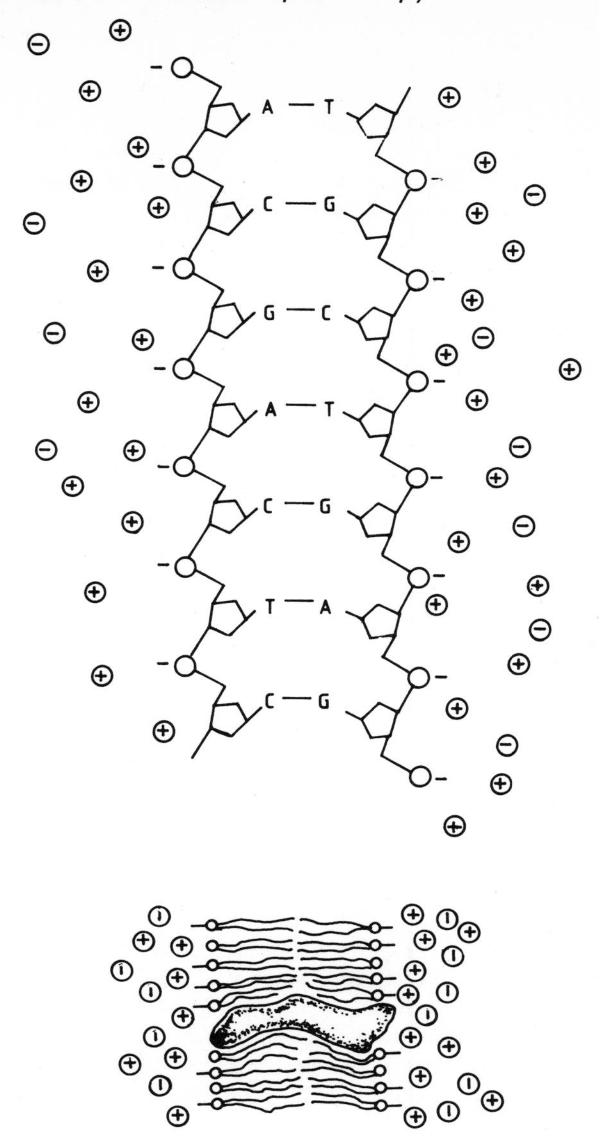

Fig. 3.8 Ionic double layers (a) surrounding the charged DNA double helix and (b) formed at charged membrane surfaces.

both cases dielectric dispersions are observed which can be attributed to the polarization of the counter-ion distribution.

Prior to the application of an electric field, the counter-ions, which are freely mobile on the surface of the macromolecule and which are free to exchange with ions in the bulk aqueous phase, show a symmetrical distribution around the macromolecule. With a field applied, there is a displacement of the surface counter-ion layer to give an asymmetrical distribution. The new equilibrium is then determined by the magnitude of the electric field and the opposing force associated with the ionic concentration diffusion gradient which tends to restore the random, symmetrical distribution.

For the model system comprising of spherical particles surrounded by a layer of counter-ions, the relaxation time is determined by surface diffusion. Schwarz has shown that the relaxation time for such a system can be described by the equation

$$\tau = \frac{a^2}{2D} \tag{3.10}$$

where a is the particle radius. D is the diffusion coefficient given, according to the Einstein relationship, by

$$D = \frac{mkT}{q}$$

where m is the surface mobility of the ion and q is the magnitude of the charge. Although equation (3.10) is in quantitative agreement with experimental results, better correlation has been achieved using analysis which takes into account ion exchange between the double layer and bulk environment and also induced electric-field effects which result in the radial polarization of the counter-ion cloud. In addition to these macroscopic tangential and radial polarizations, there may also be a contribution to the frequency response characteristics from the microscopic hopping of counter-ion charges from sites associated with the charged lipid head groups and proteins which make up the membrane surface.

In the case of rod-like molecules such as DNA, the counter-ions can be considered to be bound to the charged surface of the macromolecule and each surface charge can be represented by a potential energy well with associated resident counter-ions. In order to migrate along the charged surface under the influence of the electric field, the counter-ions must acquire sufficient energy to 'hop' over the potential energy barrier into an adjacent well. It turns out that, for macromolecules, the surface charge density is such that this energy requirement is relatively small and the counter-ion surface mobility is comparable with that of free ions in solution. The counter-ions are therefore able to respond freely to the applied electric field. The DNA system then gives rise to a frequency-dependent surface conductivity involving two separate effects. The first is the polarization of the counter-ion layer by the electric field, which takes place very quickly. The second process, the rotational relaxation of the induced dipole, is much slower and occurs at low frequencies where electrode polarization effects make observation of other dielectric loss processes difficult.

Schwarz has used a simple model to show that the polarizability and the relaxation time of such a system can be given by

$$\alpha = \frac{\alpha_o}{1 + i\omega\tau}$$

where

$$\alpha_o = 0.5(\pi\varepsilon\varepsilon_{\mathrm{eff}}\beta L^3)$$

and the relaxation time is

$$\tau = \frac{\pi\varepsilon_{\mathrm{eff}} L^2}{2mzq^2}$$

where $\varepsilon_{\mathrm{eff}}$ is the effective permittivity of the ion cloud, β is the fraction of counter-ions on the poly-ion surface (i.e. the degree of association), m is the counter-ion mobility and z is the number of counter-ions/unit length of the poly-ion, and L is the length of the molecule. In this case, the theory predicts dielectric increments which are proportional to the length and relaxation times which vary as the square of the molecular length (compared with the cube of the length for rotational relaxation).

Techniques

INTRODUCTION

Techniques for the measurement of complex permittivity over the frequency range from say 0.1mHz to 10 GHz are numerous and it is beyond the scope of this book to give comprehensive details of them all. The aim of this section is to provide a brief description of conventional a.c. bridge methods which have been employed in some form for more than forty years and to then describe in more detail the use of frequency response analysers and time domain reflectometry which are relatively new techniques.

BRIDGE METHODS

Although there are a number of different bridge configurations, the basic principle of operation is the same and involves the balancing of the impedances of the two opposite pairs of arms of the bridge. One of these arms contains the sample which may be modelled as a resistor R_s in parallel with a capacitor C_s. In the simple bridge system shown in Fig. 3.9, the bridge is balanced by reducing the voltage between the two opposite arms to zero, at which point the ratio of the two adjacent bridge arms is equal to the ratio of the impedances of the other pair of arms. So if $Z_1 = Z_2$, then the adjustable arm of the bridge (Z_v) which contains both resistive and capacitive components can be manipulated to achieve the balanced condition, and in this state Z_v is then equal to the sample impedance ($R_s = R_v$ and $C_s = C_v$).

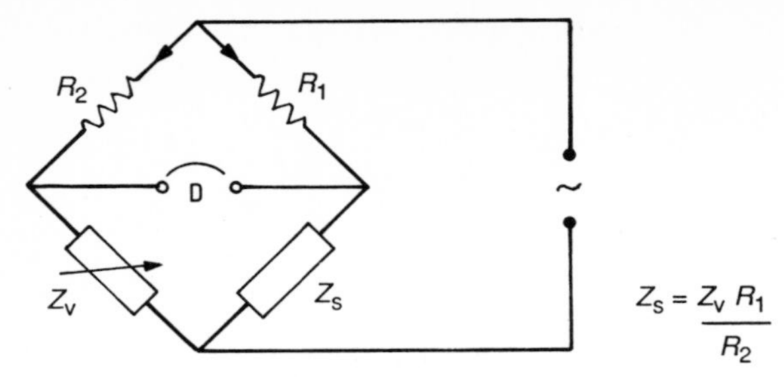

Fig. 3.9 The two-terminal a.c. impedance bridge.

In general terms, for a sample of thickness d and area A held between two parallel electrodes, the permittivity ε' and dielectric loss ε'' can be calculated from the measured values of conductance G and capacitance C using the equations

$$G=1/R=\omega\varepsilon_o\varepsilon''A/d \tag{3.11a}$$

and

$$C=\varepsilon_o\varepsilon' A/d \tag{3.11b}$$

where ω is the angular frequency ($\omega=2\pi$ frequency) and ε_o is the permittivity of free space ($\varepsilon_o=8.854\times10^{-12}$F/m). It should be noted that equation (3.11a) incorporates both frequency-dependent and steady-state (frequency-independent) contributions to the total measured conductance, G. These terms may be separated as follows:

$$G=\omega\varepsilon_o\varepsilon''_{ac}\frac{A}{d}+G_{dc}$$

where ε''_{ac} is the frequency-dependent dielectric loss and G_{dc} is the steady-state d.c. conductance.

Commercial bridges are available with frequency ranges down to 1 Hz and as high as 100 MHz. Some bridges are capable, when set-up correctly, of very accurate measurements, with resolutions in capacitance of the order of 10^{-5} pF and in conductance of $10^{-12}\,\Omega^{-1}$. A large source of error in the higher frequency region, which is particularly serious for conductive samples, is that of stray inductance. To minimize this phenomenon, a.c. bridges are constructed symmetrically so that these effects essentially cancel. However, care must be taken with sample cell design and lead arrangements in situations where inductance is likely to be a problem. Proper shielding of the bridge, leads and sample cell are also essential to eliminate stray capacitance effects. In two-terminal bridge operation, part of the capacitance measured will be due to the lead capacitance and this must therefore be subtracted to obtain the true sample capacitance. This necessity is alleviated in three-terminal operation where the low terminal is driven to ground potential. Manual bridges where the different R–C combinations are chosen by hand have now largely been replaced by automatic bridges where the R–C switching is performed under microprocessor control.

FREQUENCY RESPONSE ANALYSER

Although a.c. impedance bridges remain the most widely used instruments for dielectric measurements in the low-frequency region (sub-hertz to 100 kHz), recently another instrument, the frequency response analyser (FRA), has been shown to be capable of sensitive dielectric measurements over this frequency range. In this case, a series of sine waves of the appropriate frequency are generated and are applied to the sample. The FRA then measures the phase and amplitude of the sample response simultaneously and these parameters can be converted into permittivity and dielectric loss.

The input impedance of the analyser channels in commercially-available FRAs
ge of sample impedances that can
s range by use of high-impedance
edance from 1 MΩ to about 1 TΩ
his without change of phase and
e of the FRA. As the maximum
y limited to about 10 V, then for
measure changes of the order of
e buffer's amplifiers to very low
mploying low-drift electrometer-

mple by comparing the gain and
lance of a standard resistor and
3.10. In an automatic version the
croprocessor to give values in the
eep, it is usually necessary for the
dard resistor in order to maintain
es with frequency.
surements on the test sample, a
circuit in place of the sample and
l standard R and C combinations
oped by Pugh, eliminates the need
capacitors. Frequency response
y range as a.c. impedance bridges,
r sample cells. The same problems
ce also apply to FRA methods.
t detailed in Fig. 3.10,

Measure the Voltage
ifference (D) as a function
f frequency. Make Zv the
ample sol'n and Zs the
lank/control.
we need the cell constant A/d
Gdc will also include A/d
10 V of output from FRA will
do a lot of redox chemistry!

With the sample in circuit:

$$\frac{V_x}{V_y}=\frac{Z_s+Z_m}{Z_m}=1+\frac{Z_s}{Z_m}=1+\frac{Y_m}{Y_s}=A+iB$$

Where Z_s and Z_m and Y_s and Y_m are the impedances and admittances of the sample and standard R–C network, respectively.

With the reference capacitor in circuit:

$$\frac{V_x}{V_y'}=1+\frac{Y_m}{Y_r}=A'+iB'$$

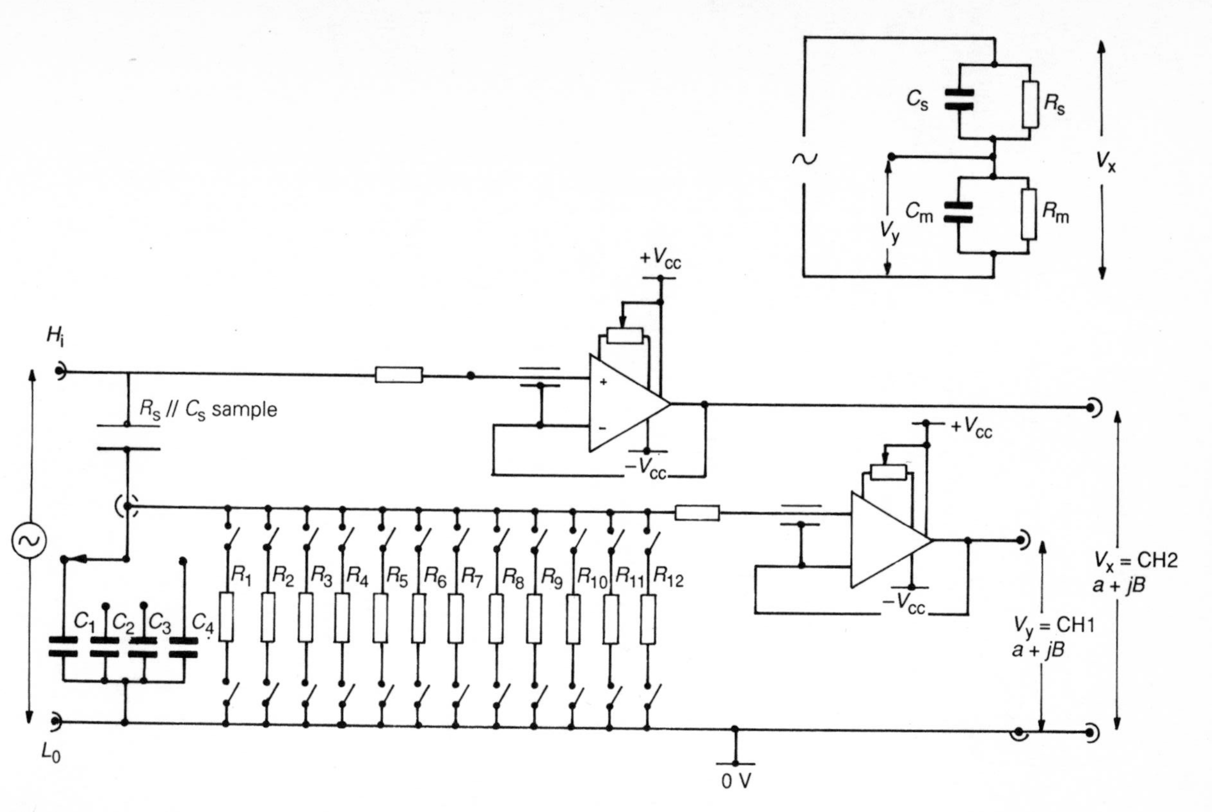

Fig. 3.10 Circuit diagram showing the principle of the frequency response analyser as used for dielectric measurements.

A and B and A' and B' are the gain and phase components of the responses made with the sample and reference air capacitor in circuit, respectively, and Y_r is the admittance of the reference air capacitor. The sample admittance may be written as $Y_s = G_s + i\omega C_s$ and by substituting for Y_m and assuming that Y_r has no conductance term, it is possible to derive the following equations for the sample conductance and capacitance:

$$G_s = \omega C_r[B(A'-1) - B'(A-1)]/[(A-1)^2 + B^2]$$
$$C_s = C_r[(A'-1)(A-1) + B'B]/[(A-1)^2 + B^2]$$

Values of G_s and C_s are dependent purely on the reference capacitor value and the FRA values measured during the two frequency sweeps and are independent of the standard resistors and capacitors employed.

TIME DOMAIN REFLECTOMETRY

In contrast to the a.c. impedance bridge and frequency response analyser techniques which involve dielectric measurements in the frequency domain, *time domain reflectometry* (TDR) is a time domain technique involving the application of a rectangular step-voltage pulse to the sample and the monitoring of the changes in the characteristics of the pulse after the reflection from a section of co-axial line filled with sample. TDR is most usefully applied to the frequency range 100 kHz to 10 GHz and has the advantage over frequency domain methods covering this part of the spectrum that a single record is sufficient to give information over a wide frequency range. TDR has undergone considerable development over the last 15 years mainly in the area of instrumentation for generating and observing very fast picosecond responses.

In a TDR experiment a train of fast-rising voltage pulses is applied to a low loss co-axial line and the waveform in the line is observed with a suitable sampling system connected to a sampling oscilloscope. The sample of interest is inserted into the line which is terminated by either a short circuit, open circuit or matched section. The applied voltage pulse and reflected sample response are combined and transformed into the frequency domain to give the desired frequency response. By choice of suitable time windows and sample cells it is possible to cover the frequency range from 100 kHz to 10 GHz.

Two TDR methods

We will now consider in more detail two TDR methods, the *direct method* and the *precision difference method.* In both cases, the sample cells take the form of open-ended, open-circuit coaxial air lines (see Fig. 3.11) positioned so as to terminate a section of low-loss co-axial cable. Care must be taken to ensure that the appropriate lengths of cable are inserted between the pulse generator and sampling head and between the sampling head and sample so as to eliminate unwanted reflections from these units from the time windows of interest. A block diagram of a typical TDR system is shown in Fig. 3.12.

Direct method In the direct method, two waveforms are required, the first defining the input step pulse and the second that of the samples' reflected response. Since it

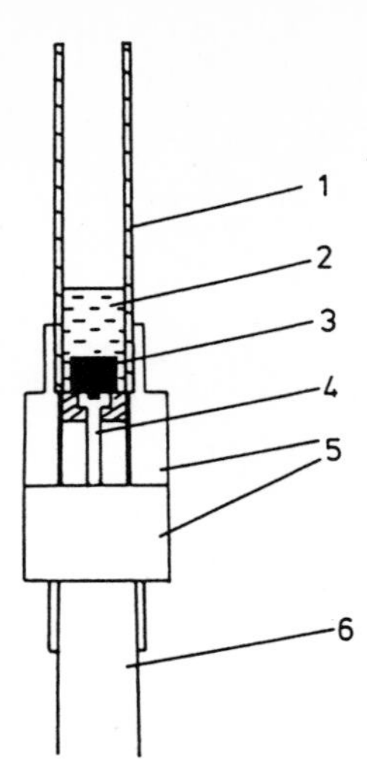

Fig. 3.11 Sample cell for high-frequency time-domain reflectometry measurements of liquids: 1, outer electrode; 2, liquid sample; 3, inner electrode; 4, connector centre pin; 5, APC7 connectors; 6, co-axial line. From Bone, S. (1988). Time-domain reflectometry: the difference method applied to conductive aqueous solutions, *Biochim. Biophys. Acta* **967**, 401–407.

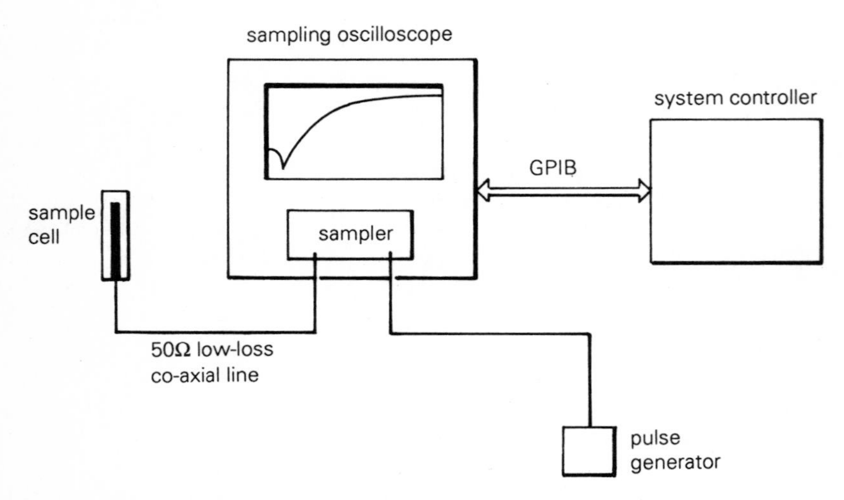

Fig. 3.12 Block diagram of the time domain reflectometry system. GPIB = General Purpose Interface Bus.

is not possible to observe the incident step pulse at the sample cell position, the nearest approximation, obtained by acquiring the reflected pulse from an empty test cell with no inner electrode, is used.

Cole and co-workers have designed a two-position cell for the measurement of the permittivity of liquids where the inner electrode can be accurately positioned in the cell for the measurement of the reflected pulse from the sample, or removed for measurement of the incident pulse, without the necessity of dismantling the test cell.

It can be shown from transmission line theory that the basic equation for determining the complex permittivity ε^* of the sample is given by

$$\varepsilon^*(\omega) = \frac{c}{d} \times \frac{v_o - r}{i\omega(v_o + r)} \times f(z) \tag{3.12}$$

where v_o and r are the Laplace transforms of the incident waveform $V(t)$ and

ll the measurement cells
or both FRA and TDR
employ two electrodes —
there will be no rejection
of interfacial redox rxns.

of the sample cell and c is the speed
cts in the sample and co-axial line
$\omega d/c)\varepsilon^{*0.5}$ and f(z) is given by

$^{*2} + \ldots$

le is usually chosen so that, for the
duces to a value of unity.
are combined to give $V - R$ and
frequency domain. Dielectric
equation (3.12). This method can
and, with additional terms to
samples. However, it is far less
method.

ifference method is most usefully
of the time domain response of a
dielectric characteristics over the
rence method has been developed
d associates for non-conducting
aqueous conducting solutions.
he reflected waveform from the
R_x together with the incident pulse
ve very accurate measurements of
ence, for instance between a dilute
he recorded waveforms and their
ethacrylate (PMMA) in benzene
ng these differences to the complex dielectric permittivities of the sample and reference is

$$\frac{\varepsilon_x^*}{\varepsilon_s^*} = \frac{(v - r_s) + (r_s - r_x)}{(v - r_s) - \left[\dfrac{i\omega d\varepsilon_s^*(r_s - r_x)}{cf(z_s)}\right]}$$

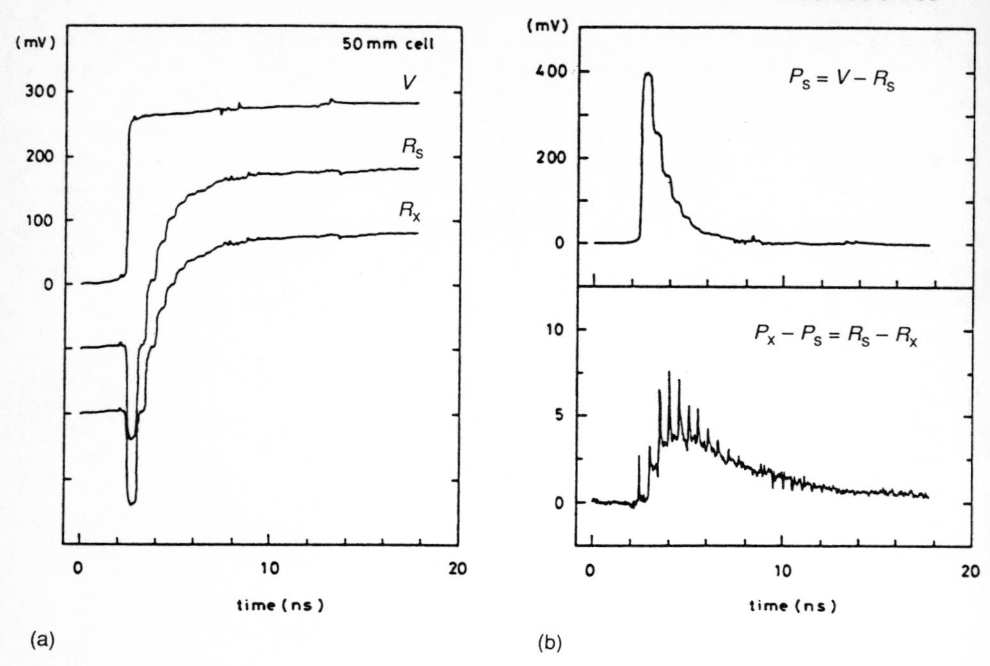

Fig. 3.13 (a) Time domain signals of V (the input signal), R_s (the reflected reference signal) and R_x (the reflected sample signal) for 4.8% w/w polymethylmethacrylate (PMMA) in benzene at 20 °C. (b) Time domain signals of $V-R_s$ and R_s-R_x for the above polymer solution.
From Nakamura, H., Mashimo, S. and Wada, A. (1982). Application of time domain reflectometry covering a wide frequency range to the dielectric study of polymer solutions, *Jap. J. Appl. Phys.* **21**(3), 467–474.

where ε_s^* is the complex permittivity of the reference dielectric and v_o, r_s and r_x are the Laplace transforms of V, R_s and R_x, respectively. For conductive samples such as solutions of biological molecules, additional terms are required to take account of the effective d.c. conductivity of the solutions.

Transformations from the time into the frequency domain are usually carried out using the Samulon modification of the Shannon sampling theorem, using equations of the form

$$i\omega\, f(\omega) = (\theta/\sin\theta)\exp(i\theta)\sum_{n}^{N}\exp(-i\omega n\Delta)[F(n\Delta)-F(n\Delta-\Delta)] \qquad (3.13)$$

where $\theta = \omega\Delta/2$ and Δ is the time interval between consecutive data points. It is usually arranged that the frequency range of interest is such that

$$\theta/\sin\theta = 1 \qquad \text{and} \qquad \exp(i\theta) \simeq 1+\theta$$

in which case equation (3.13) may be written as

$$i\omega\,\mathrm{f}(\omega)=[1+\theta]\sum_{n}^{N}[\cos(\omega n\Delta)-i\sin(\omega n\Delta)][F(n\Delta)-F(n\Delta-\Delta)]$$

Experimentally it is necessary to sample the required waveforms over a sufficiently wide time window to capture enough of the waveform to adequately characterize the response and to avoid truncation errors. This requirement is fulfilled by ensuring that the time window extends to at least five times the relaxation time of the dielectric sample.

The single largest source of systematic error in all TDR methods is time shift errors. These arise mainly from drift of the tunnel diode pulse relative to the scanning sweep and, if uncorrected, result in a time mismatch between individual traces. Errors of this type become increasingly serious at high frequencies and, without adequate time referencing, can produce erroneous permittivity data above 100 MHz. Various methods have been employed to reduce timing errors. The most commonly used time referencing technique involves extrapolation of the initially rising portions of the incident and reflected waveforms to the point where they cross the baseline. This then gives a common time reference point which can be used to align the appropriate waveforms. A method which produces better results is one where a marker impulse is used as a timing reference. The incident pulse is split into two, one pulse being used to characterize the sample response while the other carries the marker. Very accurate time referencing can be achieved using this method but an oscilloscope with two synchronized sampling channels is required.

ELECTRODE POLARIZATION

The frequency range over which meaningful dielectric data can be obtained for aqueous solutions of biological materials is limited by a phenomenon known as *electrode polarization*. Electrode polarization is the term used to describe the accumulation of ionic species at the boundary between the electrode and aqueous sample and is particularly marked at low frequencies and for conductive samples. The build up of ions occurs as a result of the inability of the electrodes to compensate the ionic charges at a rate comparable with their arrival at the electrode surface. Thus at frequencies typically <1 MHz for biological solutions, ionic layers are formed at this interface which result in an increase in the measured capacitance. The effect of electrode polarization on the capacitance and conductance of a blood sample can be seen in Figs. 3.14 and 3.15. The conductance shows a marked fall and the capacitance a considerable increase with decreasing frequency, as electrode polarization becomes increasingly important. If the electrode polarization and sample contributions are modelled by the circuit illustrated in Fig. 3.16, then following simple circuit theory used by Schwan, the equations relating these to the measured resistance and capacitance are

$$R=R_s+R_p+(R\omega C)^2R_s$$
$$C=C_s+1/(\omega^2R^2C_p)$$

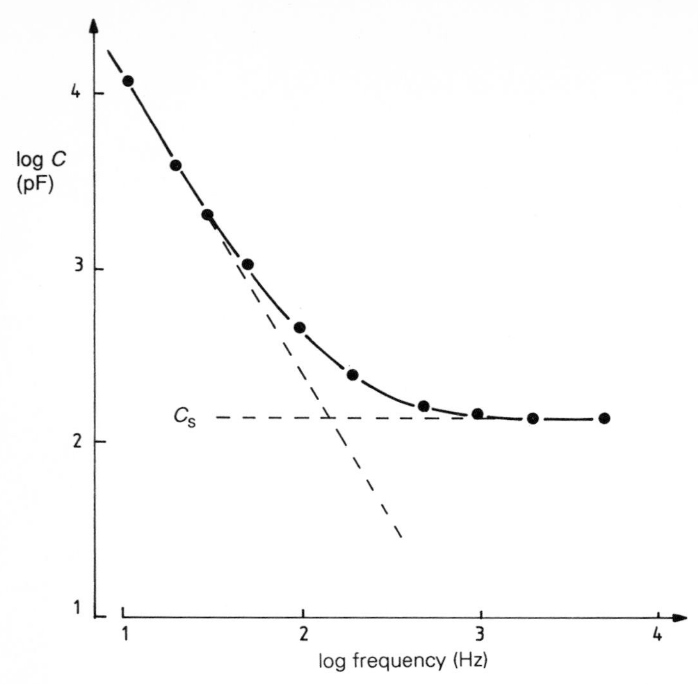

Fig. 3.14 Frequency dependence of the total measured capacitance C of blood in a sample cell exhibiting low electrode polarization. The measured capacitance C is the sum of the true sample capacitance and the contribution from the electrode polarization term $(1/\omega^2 R^2 C_p)$.
Data after Schwan, H.P. (1966). *Biophysik* **3**, 181–201.

where R_s and R_p and C_s and C_p are the resistances and capacitances associated with the sample and electrode polarization respectively, and R and C are the resistance and capacitance for the equivalent parallel R–C combination.

A number of different methods have been employed to correct for the electrode polarization contribution to the total observed impedance. One technique is to use a *sample cell where the electrode separation can be varied.* With small electrode separations the electrode polarization will be the dominant effect whilst at larger inter-electrode distances the sample contribution should be relatively larger. Care must be taken, when using this method, to minimize stray capacitances which become increasingly important for larger electrode spacings. By performing dielectric measurements at various inter-electrode distances it is possible, providing the sample contribution is significant, to separate the electrode polarization from the sample polarization.

Another method, known as the *substitution method*, involves the evaluation of the electrode polarization contribution by making dielectric measurements on an

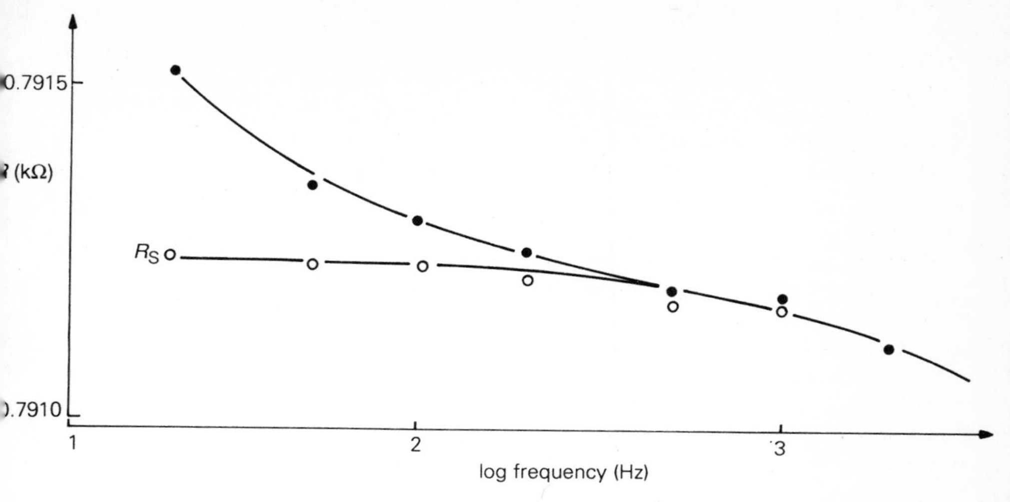

Fig. 3.15 Frequency dependence of the resistance of blood measured in a sample cell of small electrode polarization; R is the total measured resistance, R_s is the true sample resistance.
Data from Schwan, H.P. (1966). *Biophysik* **3**, 181–201.

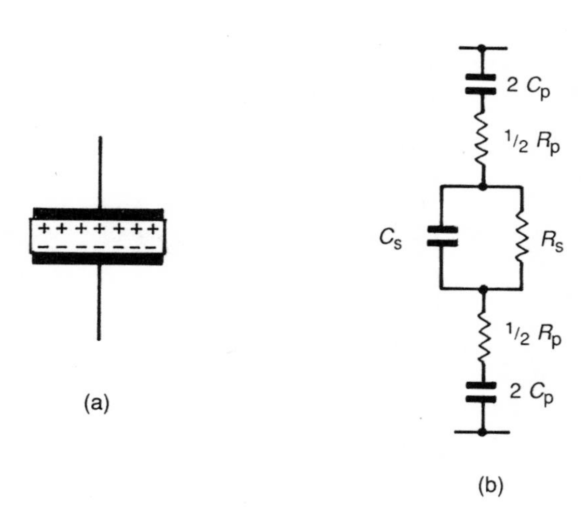

Fig. 3.16 (a) Dielectric sample cell containing electrolyte solution exhibiting electrode polarization at the electrolyte/electrode interface. (b) Equivalent circuit showing electrode polarization resistance R_p and capacitance C_p and bulk sample resistance and capacitance R_s and C_s.

electrolyte solution of the same conductivity as that of the sample solution. The measured electrode polarization can then be subtracted to give the dielectric response of the sample.

Experimental methods for reducing the magnitude of the electrode polarization involve various electrode treatments which have the effect of increasing the effective electrode area accessible to ions. This may be achieved by roughening the electrode surface but far better results can be obtained when platinum electrodes are used onto which a layer of platinum black has been electro-deposited. This method cannot however be used with certain biological materials. A number of proteins, for example, are chemi-absorbed by platinum black electrodes and this effect can itself produce erroneous dielectric data.

A method known as the *four-electrode technique* has also been employed for low-frequency measurements in the absence of electrode polarization. In this technique, two electrodes generate the a.c. signal whilst two other electrodes probe the potential across the sample. Provided that the potential electrodes do not draw any current, electrode polarization should not be present.

Selected reading

Bone, S. (1988). Time domain reflectometry: the difference method applied to conductive aqueous solutions, *Biochim. Biophys. Acta*, 401–407.

Cole, R.H. (1977). Time domain reflectometry, *Ann. Rev. Phys. Chem.* **28**, 283–300.

Cole, R.H., Mashimo, S. and Winsor, P. IV (1980). Evaluation of dielectric behaviour by time domain spectroscopy. 3. Precision difference method, *J. Phys. Chem.* **84**, 786–793.

Daniel, V.V. (1967). *Dielectric Relaxation.* London, Academic Press.

Davies, M. (1965). *Some Electrical and Optical Aspects of Molecular Behaviour.* Oxford, Pergamon Press.

Grant, E.H., Sheppard, R.J. and South, G.P. (1978). *Dielectric Behaviour of Molecules in Solution.* Oxford, Clarendon Press.

Nakamura, H., Mashimo, S. and Wada, A. (1982). Application of time domain reflectometry covering a wide frequency range to the dielectric study of polymer solutions, *Jap. J. Appl. Phys.* **21**, 467–474.

Pugh, J. and Ryan, J.T. (1979). Automated digital dielectric measurements, in *IEE Conference Proceedings on Dielectric Materials, Measurements and Applications* **177**, 404–407.

Pugh, J. (1984). Dielectric measurements using frequency response analysers, in *IEE Conference Proceedings on Dielectric Materials, Measurements and Applications* **239**, 247–250.

Schwan, H.P. (1966). Alternating current electrode polarisation, *Biophysik* **3**, 181–201.

Schwarz, G. (1972). Dielectric polarisation phenomena in biomolecular systems, in *Dielectric and Related Molecular Processes*, Vol. 1. The Chemical Society Specialist Periodical Reports.

Chapter 4

Dielectric and Conduction Properties of Biomolecules

Over the past fifty years dielectric and conduction properties of biological molecules have been the subject of many investigations. As a result there exists a wealth of valuable information on the charge transport and rotational properties of many biomolecules. Amino acids, proteins and nucleic acids, lipids, cell and liposome suspensions and tissues have been characterized over a wide frequency spectrum ranging from a few hertz to tens of Gigahertz. In certain cases, dielectric measurements have been used to probe the physical changes which take place in biologically important structures, for example in the lipid-phase transition process in membranes. Dielectric spectroscopy has also enabled investigations into the possible harmful effects of electromagnetic fields on biomolecular systems. Little is known about how environmental electric fields may perturb the subtle balance of electric fields existing within cells, for example the relatively large electric fields which are maintained across cytoplasmic, mitochondrial and nuclear membranes. In this chapter the inherent charge distribution and the electrical responses of a wide range of biological systems from relatively simple amino acid solutions to very complex animal tissues will be discussed. It will be seen that in some cases molecular dipoles determine the electrical characteristics whilst in others induced polarization effects are responsible for the observed phenomena.

Amino acids

Amino acids are the basic structural units from which proteins are formed. They have the general structure

$$\begin{array}{c} NH_3^+ \\ | \\ H—C—COO^- \\ | \\ R \end{array}$$

where the R group is known as the side chain. Amino acids can exist in two optically active forms, the D-isomer and the L-isomer, although only the L-isomer occurs in protein structures. Depending on the pH of the local environment, amino acids can take on a number of different ionized states, as described by Fig. 4.1.

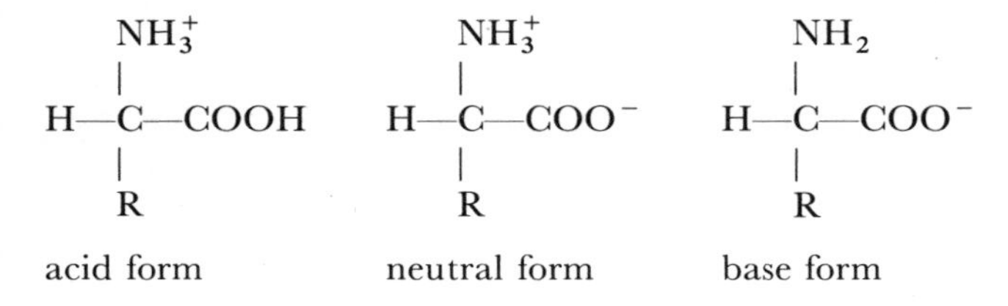

Fig. 4.1 The pH-dependent ionic forms of the α-amino acids.

Amino acids at acidic pH values possess protonated amino groups whilst at alkaline pH the carboxyl group alone carries a charge. At neutral pH values the dipolar (or zwitterionic) form predominates in which both amino and carboxyl groups are ionized.

Each amino acid possesses a different side chain and these can be small, for example in the case of alanine ($R=CH_3$), or relatively large, for example in the case of arginine ($R=(CH_2)_3NHCNH_2^+NH_2$). Each side chain also has a particular charge distribution. Some amino acids can be designated non-polar such as glycine, alanine and leucine where there is little or no charge asymmetry associated with the side chain, others are ionized at physiological pH and therefore are strongly polar such as glutamic acid (negatively charged) and lysine (positively charged). Others, whilst not being ionized at neutral pH, still possess a degree of charge asymmetry and are weakly polar. Examples in this category are glutamine and serine. There are twenty common amino acids whose specific properties are governed by the nature of the side chain and these are grouped according to side-chain polarizability in Table 4.1. The twenty side-chain variations are responsible for the diverse structure and function of the whole range of proteins found in all living systems.

Solutions of amino acids generally exhibit two dielectric dispersions in the gigahertz frequency region. The lower frequency dielectric loss is associated with the amino acid whilst bulk water is responsible for the dispersion at higher frequencies (the relaxation frequency for water is 17 GHz at 25°C). The relaxation frequencies of the bulk water and amino acid dispersions are usually close and it is

Table 4.1 The α-amino acids

Non-polar side chains	Polar side chains	Ionised side chains
Glycine	Serine	Aspartic Acid
Alanine	Threonine	Glutamic Acid
Valine	Tyrosine	Histidine
Proline	Tryptophan	Lysine
Leucine	Glutamine	Arginine
Isoleucine	Cysteine	
Phenylalanine	Asparagine	
Methionine		

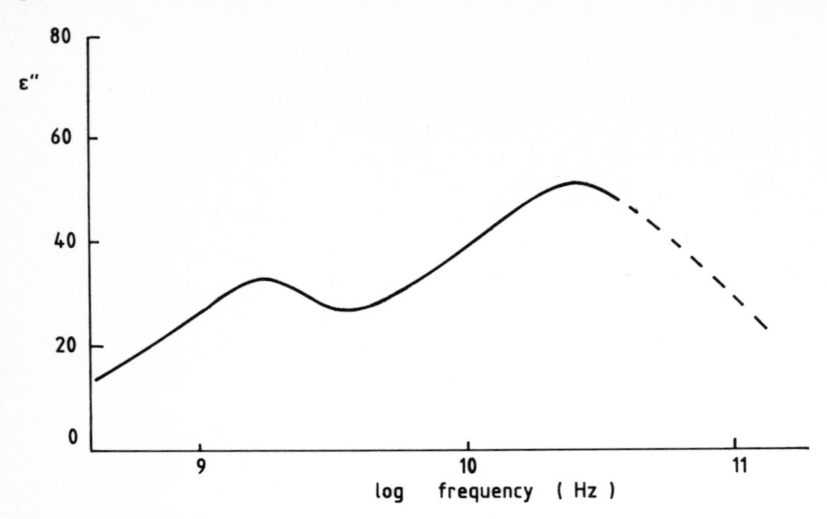

Fig. 4.2 Dielectric spectra of 1 M aqueous solution of α-alanine at pH 7.1.

often not possible to observe well-separated dispersion regions, the amino acid peak being partially buried by the bulk water dispersion as depicted in Fig. 4.2 for alanine.

As a result of their ionic character, amino acids in neutral solution possess unusually large dipole moments, larger in fact than that of water ($\mu = 1.84$ Debye units). This results in an increase in the value of the permittivity when amino acids are dissolved in aqueous solvent since the solvent of lower permittivity per unit volume is replaced by a solute of higher permittivity per unit volume.

There is a considerable amount of evidence to suggest that the origin of the observed lower frequency dielectric dispersion is associated with the rotation of the ionized amino-carboxyl dipole which is a characteristic of the zwitterionic form of amino acids. A simple calculation of the zwitterionic dipole moment from the product of the charge and the distance separating the amino and carboxyl groups (0.32 nm) produces a value of 15.3 Debye. This value compares favourably with those from a number of studies on α-amino acids which report values for the effective dipole moment of about 19 Debye. In the case of amino acids where the charge separation is greater such as for β-alanine and ε-aminocaproic acid (shown

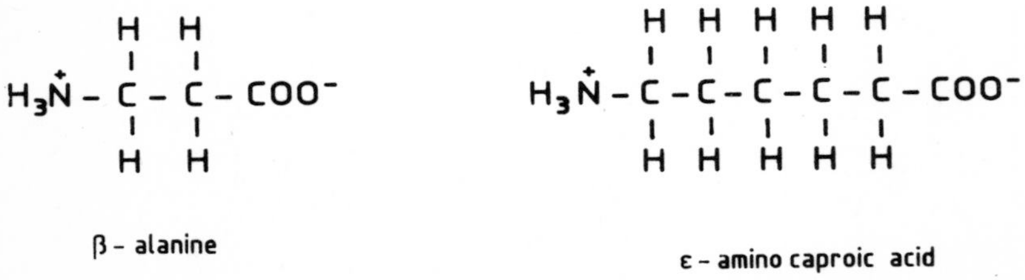

$$H_3\overset{+}{N}-\underset{H}{\overset{H}{C}}-\underset{H}{\overset{H}{C}}-COO^-$$

β - alanine

$$H_3\overset{+}{N}-\underset{H}{\overset{H}{C}}-\underset{H}{\overset{H}{C}}-\underset{H}{\overset{H}{C}}-\underset{H}{\overset{H}{C}}-\underset{H}{\overset{H}{C}}-COO^-$$

ε - amino caproic acid

Fig. 4.3 (a) β-alanine; (b) ε-aminocaproic acid.

Table 4.2 Dielectric parameters of some amino acids

Amino acid	*Specific increment* ($\Delta\varepsilon/c$)	μ_{eff} (*Debye*)	MW	τ (*picoseconds*)
Asparagine	0.20	17.3	132.1	
Aspartic acid	0.28	20.1	133.1	
Glutamic acid	0.26	19.5	147.1	
Glutamine	0.21	19.8	146.1	
Leucine	0.25	19.1	131.2	
Glycine	0.23	18.1	75.1	49
Proline	0.29	20.6	151.1	69
α-alanine	0.23	18.3	89.1	92
β-alanine	0.34	22.4	89.1	87
ε-aminocaproic acid	0.59	29.3	131	129

in Fig. 4.3), the measured dielectric increment and derived effective dipole moment are proportionately larger. This type of dependence of the dipole moment on the separation distance of the two charges has also been observed in a dielectric study on glycine polypeptide solutions where a linear relationship has been shown to exist between the dielectric increment and the number of bonds separating the amino and carboxyl groups.

Further evidence for an interpretation in terms of a rotation of the amino acid molecule comes from the molecular-weight dependence of the relaxation time. A fairly linear relationship has been observed between the molecular weight of α-amino acids and their observed relaxation times (see Table 4.2). A parameter which similarly affects the activation energy of the rotational relaxation process is the solvent viscosity, as discussed in Chapter 3. In studies of the dielectric properties of the amino acid proline dissolved in a number of different solvents, the relaxation time was found to be dependent upon the solvent viscosity according to equation (3.7) (p. 44). A linear relationship has also been observed between solution viscosity and the product of the absolute temperature and dielectric relaxation time for solutions of triglycine. These results are consistent with a scheme involving the rotation of the amino acid in the electric field.

In solution, the charged carboxyl and amino groups interact strongly with the surrounding polar water molecules. This electrostatic interaction effectively induces a dipole moment in addition to that attributable to the amino acid in isolation. A number of water molecules are, in effect, hydrogen bonded to the amino acid and this results in the polarized charge distribution extending into the aqueous medium as shown in Fig. 4.4. In practice the extent of the solvent-induced dipole moment is relatively small compared with the large amino acid moment. Typically the induced dipole moment accounts for some 10% of the measured dielectric increment.

Onsager and Kirkwood have developed theories to take account of electrostatic interactions in the analysis of dielectric data. The effective (measured) dipole

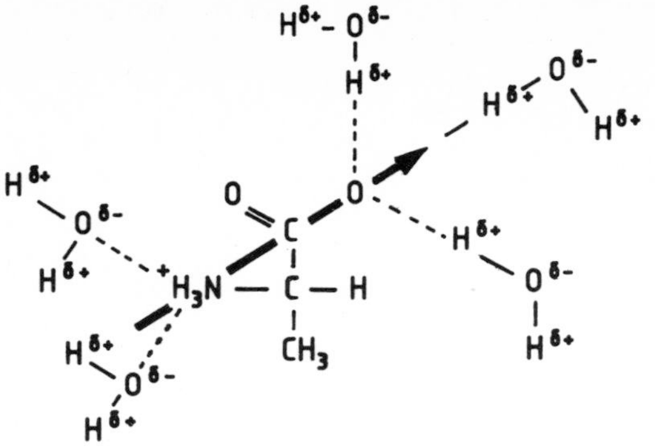

Fig. 4.4 Electrostatic interaction between the charged zwitterionic groups of an amino acid and polar water molecules.

moment, μ_{eff}, can be related to the actual dipole moment, μ_i, of the isolated amino acid by the equation

$$\mu_{eff} = g^{0.5}\mu_i$$

where the quantity g is known as the Kirkwood correlation factor which accounts for short-range electrostatic interactions between relaxing molecules. For zero correlation, the parameter g reduces to unity; for pure water which is capable of hydrogen bonding a maximum of four other water molecules, g has the value 2.8. Unfortunately, without a precise knowledge of the aqueous environment in the immediate vicinity of the amino acid, it is not possible to derive an accurate value of g. It is usual therefore to present amino acid dipole moments in terms of the effective (measured) dipole moment $g^{0.5}\mu_i$. Values of the effective dipole moment for a number of amino acids are shown in Table 4.2.

As we have seen, the carboxyl and amino zwitterions may not be the only charged groups in the amino acid structure. There are a number of amino acids which also possess a side chain which is ionizable. In this case one might imagine the dipole moment as being due to the vector addition of the contributions from both the side chain and the zwitterionic charges.

So far we have looked at the dielectric response of solutions of amino acids at pHs close to the isoelectric points, where the zwitterionic form is by far the most prevalent. At pH values on either side of the isoelectric point (pI) value, a significant fall in the polarizability and dielectric increment is observed as the zwitterionic form is replaced by the single positively- and negatively-charged species present at higher and lower pH respectively (Fig. 4.1). The trend for glycine is shown in Fig. 4.5.

The discussion of the dielectric properties of amino acids has centred around the response of the zwitterionic amino-carboxyl dipole since this is by far the dominant entity. When amino acids link together to form polypeptides and proteins, the linking peptide bond is formed from a condensation reaction involving the α-

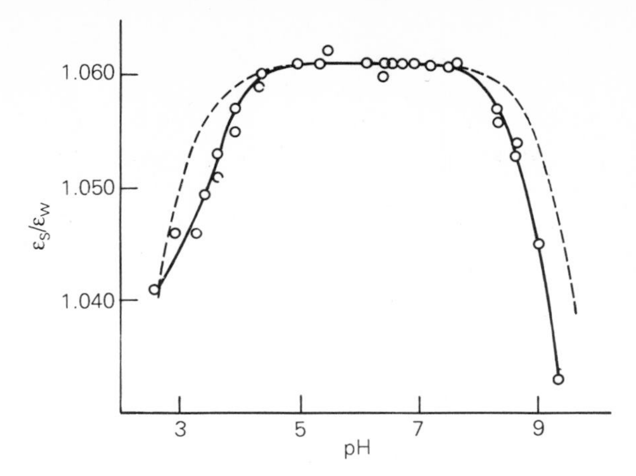

Fig. 4.5 Variation of the low-frequency dielectric constant of glycine (ε_s) relative to that of water (ε_w) with pH. The dashed line indicates the characteristic expected using accepted values of dissociation constants.
From Dunning, W.J. and Shutt, W.J. (1938). The dielectric constants of zwitterionic ions and polar molecules as related to pH, *Trans. Faraday Soc.* **34**, 479–485.

amino and carboxyl groups. Although the peptide bond is also polar, it is motionally restricted and therefore – as an individually-relaxing entity – exhibits a relatively small effective dipole moment. We will see in the next section how the peptide bond is an important factor in determining the total dipole moment of the protein, but how the contributions from polar side groups and polar ligands such as bound water must also be considered.

Proteins

Proteins are crucial to virtually every biological process. As enzymes, they catalyze innumerable chemical reactions which would otherwise occur imperceptibly slowly. They are active as carrier and storage molecules, in muscle contraction and in mechanical support. As antibodies, they are responsible for immune protection and as receptors in the nervous system for the generation and transmission of nerve impulses.

Proteins are polymers built up from amino acids. Their great diversity and versatility is derived from the properties of the twenty different amino acid side chains that may be present. The protein has an unbranched structure and is formed by condensation reactions which progressively add amino acid units to the polypeptide chain. Each monomer unit is connected by a peptide bond and possesses the side chain of the amino acid from which it was formed.

As the polymerization process progresses, the polypeptide chain folds in a way

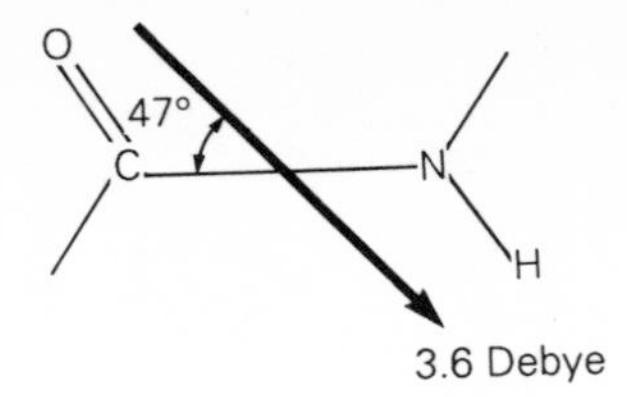

Fig. 4.6 The magnitude and direction of the dipole moment of the peptide unit.

which is under the direction of the side-chain sequence into the three-dimensional structure which characterizes the active protein. A variety of intramolecular interactions influence the final protein conformation: electrostatic and torsional interactions, dispersion forces and solvent interaction.

From a dielectric point of view, the protein molecule is a conglomeration of interacting dipoles of various magnitudes and alignments. Each peptide unit possesses a dipole moment (Fig. 4.6) and the protein can therefore be considered to be a string of connected dipoles. The side chains which possess a charged or polar character also contribute to the total dipole moment of the protein. In order to achieve the lowest energy configuration the protein tends to fold in such a way as to allow the ionized and polar side chains to be in contact with the aqueous medium and the non-polar side chains to be buried within the protein interior. So, as a general rule, most of the charged and polar groups are found on the protein surface. The conformation of the protein and the way in which the individual dipoles are aligned with respect to one another in the three-dimensional structure determines how the dipoles interact. The resultant dipole moment of the protein molecule as a whole involves the vector addition of all of the constituent dipoles, as illustrated in Fig. 4.7. An example of a situation where the dipoles associated with the peptide bonds are totally additive is the extended α-helix where these dipoles line up along the entire length of the helix. The α-helix contains 3.65 peptide units per helix turn with each peptide unit aligned approximately in the same direction as illustrated in Fig. 4.8, giving a dipole moment/helix turn of 13.1 Debye units. Small regions of α-helix in a protein can, therefore, make a relatively large contribution to the protein's total dipole moment. In practice, however, given the size of the protein molecules and the number and magnitude of constituent dipoles involved, the resultant dipole moment is not as large as might be expected. This suggests that there is a high degree of electrostatic compensation within the protein structure effectively resulting in cancellation of constituent dipole moments.

It should be understood that in the discussion so far proteins have been treated as static structures with fixed permanent dipole moments. This time-averaged structure is, of course, a simplification. Proteins are dynamic molecules exhibiting a whole range of individual, collective and cooperative types of internal motion which scale many orders of magnitude of amplitude, energy and time. The importance of the dynamic picture of the protein molecule will be discussed later.

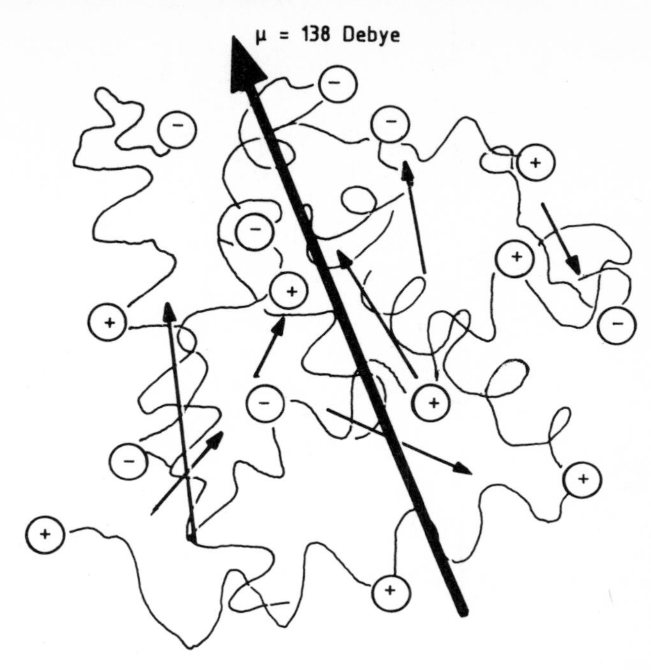

Fig. 4.7 Schematic diagram of a globular protein showing local structural dipoles due to, for example, α-helical sections (small arrows) and the total permanent dipole moment of the protein as a whole (large arrow).

Globular proteins in solution exhibit a dielectric dispersion, known as the β-dispersion, centred in the low Megahertz region of the frequency spectrum. As with amino acids, proteins possess larger dipole moments/unit volume than bulk water and this results in protein solutions exhibiting larger low-frequency permittivities than bulk water. Also, in a way similar to the amino acids, charged groups on the protein surface are involved in electrostatic interactions with the immediate aqueous environment.

The way in which the real and imaginary parts of the permittivity vary over the frequency range 100 KHz to 20 GHz for a typical globular protein in solution is shown in Fig. 4.9. The magnitude of the β-dispersion for a particular protein is determined by the concentration of the protein in solution and by the magnitude of the resultant dipole moment according to equation (3.8) (p. 44). This equation may be used to calculate the effective dipole moment of the protein molecule from the magnitude of the dielectric increment. Table 4.3 lists the dielectric increments and derived dipole moments for several proteins.

Although there remains some uncertainty as to the origin of the β-dispersion, a detailed study of the possible mechanisms involved by South and Grant has led to a generally accepted scheme involving the rotation of the protein molecule in the

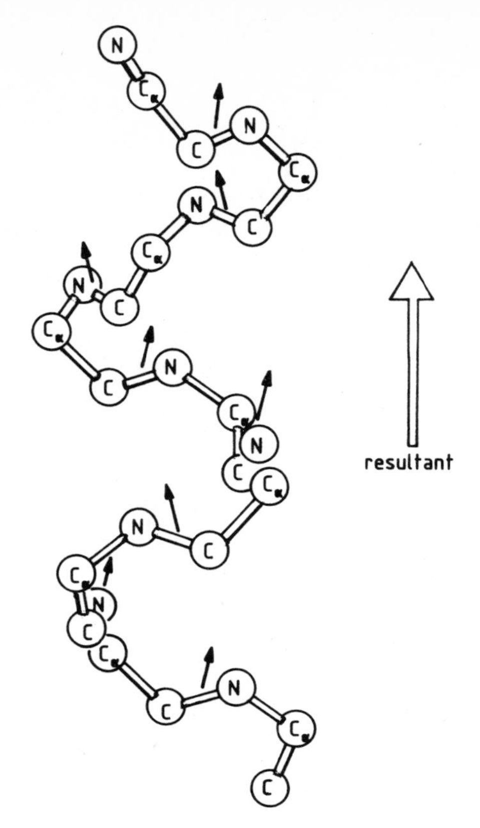

Fig. 4.8 The dipole moment of the α-helix showing the additive nature of individual peptide dipoles.

electric field. Evidence for such an interpretation, although not conclusive, has come from the finding that the relaxation time of the β-dispersion is proportional to the viscosity of the solution according to equation (3.7). It is envisaged that the protein plus its water of hydration rotate as one unit in the low megahertz frequency region. Other mechanisms which may influence the dielectric properties of protein solutions involve proton and counter-ion fluctuations on the protein surface.

Recent analysis of the β-dispersion exhibited by bovine serum albumin has indicated the existence of two separate loss processes and these have been attributed to the rotation of the ellipsoidal protein molecules about their a- and b-axes. Two characteristic relaxation times are predicted as described in Chapter 3, involving the rotation of the a-axis about the b-axis, and the rotation of the b-axis about the a-axis.

In addition to the β-dispersion shown centred at 1 MHz, two further dielectric

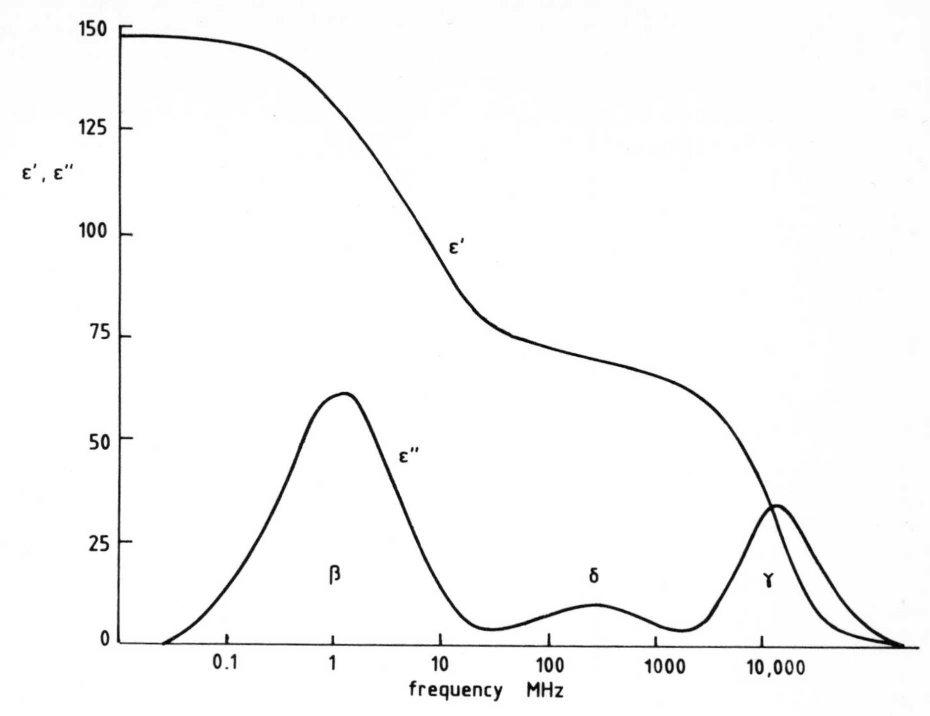

Fig. 4.9 The complex dielectric spectra of a typical globular protein in solution.

Table 4.3 Dielectric increments and derived dipole moments of some globular proteins

Protein	*Dielectric increment*, $\Delta\varepsilon$ (m^3/kg)	*Dipole moment*, μ (*Debye*)
Myoglobin (horse heart)	0.12	143
Haemoglobin (bovine)	0.32	465
Albumin (egg)*	0.10	250
Albumin (horse serum)*	0.17	380

*Data from Oncley, J.L. (1943). In *Proteins, Amino Acids and Peptides*, Eds. Cohn, E.J. and Edsall, J.T., Chapter 22. New York, Reinhold.

losses are apparent. The larger of the two, the γ-dispersion, centred at 17 GHz describes the relaxation of bulk water molecules. The second much smaller loss peak, the δ-dispersion, extends over a relatively wide frequency range. Detailed analysis of this dispersion has shown it to be composed of two separate dielectric losses. These dispersions have been attributed to the orientation of polar side groups and peptide bonds and also to the rotation of water molecules bound to the hydrophilic sites of the protein molecule. These bound water molecules are rotationally hindered through being hydrogen bonded to the protein molecule and therefore exhibit substantially longer relaxation times than normal bulk water molecules. As we shall see later, bound water molecules may play an important role in facilitating the biological activity of enzymes which are an important class of globular protein.

DNA

Deoxyribonucleic acid (DNA) is the molecule in which genetic information is stored. DNA is a very long linear macromolecule consisting of two anti-parallel helices. Both helical backbones are formed from sugar–phosphate units in which the deoxyribose sugar moieties are linked by phosphodiester bridges as shown in Fig. 4.10. DNA also contains four bases: two purines (adenine and guanine) and two pyrimidines (thymine and cytosine). These are bound to the sugars and engage in multiple hydrogen-bond interactions with the appropriate bases associated with the facing helix. The sequence of bases in the DNA structure codes for the sequence of amino acids in the structure of a protein. When transposed onto an RNA template, each set of three bases codes for a single amino acid.

At physiological pH, the phosphate groups in the DNA backbone are ionized and the sugar–phosphate units possess relatively large dipole moments. These dipoles are oriented in essentially the same direction along the entire helix and in a single helical molecule these dipole moments would be additive and produce a large resultant dipole moment. In the case of the DNA molecule, however, where there are two anti-parallel helical chains, the two sugar–phosphate dipoles are in opposition and cancel (Fig. 4.10) giving a net sugar–phosphate dipole moment for the whole molecule of zero. Although there will be a molecular dipole moment resulting from the vector addition of the contributions from the base pairs, this is not large and is not observed in the dielectric spectra of DNA solutions.

Solutions of DNA exhibit at least two dielectric loss phenomena in addition to the usual loss associated with normal bulk water. The lowest frequency relaxation, with a relaxation time of between 1 ms and 1 μs depending on the chain length, is known as the α-dispersion and is characterized by a dielectric increment of typically 1000. In this case, the process involved is not the orientation of a permanent dipole moment, but rather counter-ion redistributions which produce induced dipole moments. Positively-charged metal ions are weakly bound to the negatively-charged phosphate groups of the DNA and are in dynamic equilibrium with ions in the immediate vicinity of the DNA molecule. Studies of the dielectric

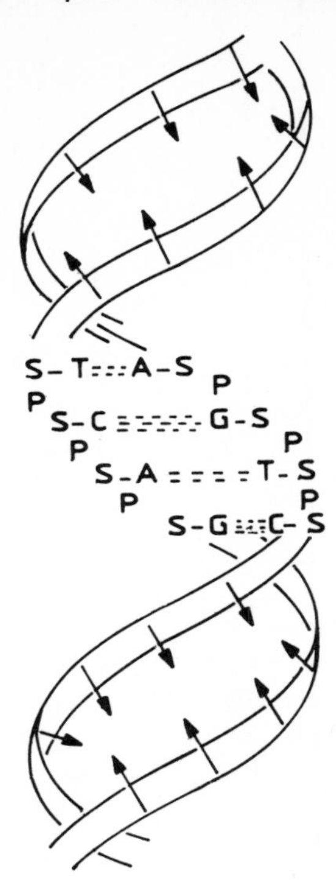

Fig. 4.10 The DNA double helix showing the anti-parallel nature of the sugar–phosphate dipoles. S = sugar; T = thymine; C = cytosine; A = adenine; G = guanine; P = phosphate.

properties of solutions of DNA with different counter-ion types have shown that the lower frequency dispersion can be attributed to an electric field-induced dipole moment originating from the redistribution of associated counter-ions around the DNA molecule, as described in Chapter 3. The application of an electric field slightly perturbs the equilibrium distribution of the counter-ions along the DNA molecule and results in the induction of a dipole whose value is roughly proportional to the square of the length of the molecule. This behaviour is typical of polyelectrolytes in solution and the dielectric properties of DNA have been found to be similar in many respects to those of polyelectrolytes.

The dielectric spectra of a dilute solution of DNA is shown in Fig. 4.11. The small dispersion centred around 100 MHz is now generally thought to originate from the motions of the polar bases and phosphate groups re-orienting as individual dipoles in the electric field. An alternative proposed mechanism is one involving short-time fluctuations of the counter-ions over a limited number of sugar–phosphate units, the extent of these fluctuations being determined by local potential barriers. The dielectric loss phenomena observed in this case can be

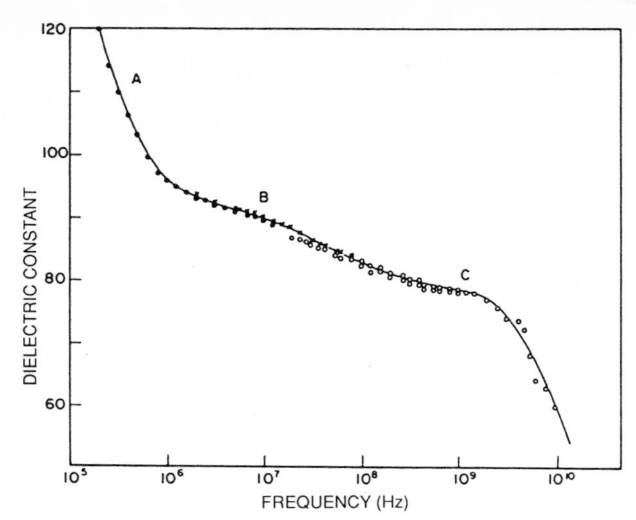

Fig. 4.11 Frequency dependence of the permittivity of 1% DNA solution. A = measurements obtained using an impedance analyser, B = using an RX meter, C = using a time domain reflectometry system.
From Takashima, S., Gabriel, C., Sheppard, R.J. and Grant, E.H. (1984). Dielectric behaviour of DNA solution at radio and microwave frequencies, *Biophys. J.* **46**, 29–34. Reproduced with copyright permission from The Biophysical Society.

described in terms of a *charge-carrier hopping model*. In this scheme, the charged counter-ions and phosphate groups constitute the charge carriers and the trapping sites, respectively. This model predicts a maximum in the polarizability of the molecule when there are equal numbers of charge carriers and trapping sites. For increasing ionic concentrations, theory predicts an initial rise in permittivity to the point at which charge carriers and trapping sites are available in equal numbers followed by a fall as the number of sites become saturated. This phenomenon has been observed, at least qualitatively, in experimental studies on DNA solutions.

Recently, both theoretical and experimental studies on specific short-chain DNA and plasmid (circular) DNA solutions indicated the existence of resonance absorptions at microwave frequencies. It was envisaged that microwave energy could be coupled to the DNA structure exciting internal acoustic modes of the molecule and setting up standing wave motions. Although resonance absorption was not detected in long-chain DNA which would be of biological relevance, in nature the DNA molecule is extensively interwoven with proteins and short resonant lengths may therefore exist between protein sites. The possibility that microwave energy might influence fundamental biochemical processes would obviously have far-reaching implications. However, some recent detailed dielectric studies have failed to detect resonant absorptions in plasmid DNA and so there must remain some doubt as to their existence.

Lipids

Lipids are a major constituent of membranes and are amphipathic, i.e. they contain both hydrophilic and hydrophobic moieties in the form of polar head groups and non-polar tails. Phospholipids, the most abundant form, are derived from either glycerol, a C_3 alcohol or sphingosine, a more complex alcohol. Phosphoglycerides, the phospholipids derived from glycerol, consist of a glycerol backbone, two fatty acid chains and a phosphorylated alcohol as shown in Fig. 4.12(a). The structure of 1-palmitoyl-2-oleoylphosphatidylcholine, a common phosphoglyceride found in most membranes of higher organisms, is shown in Fig. 4.12(b).

Although lipids may be investigated as individual molecules using, for example, chloroform as solvent, the majority of dielectric studies have concentrated on their properties in membrane systems. Membranes separate the cell from, and mediate communication with, the cell environment. They are sheet-like structures consisting mainly of lipids and proteins. The lipids conform to a bilayer structure (Fig. 4.13) of thickness between 60 and 100 Å with hydrophilic head groups facing outwards into the aqueous environment and into the cell interior. The lipid bilayers are cooperative structures held together in the main by the hydrophobic interaction which minimizes the number of hydrocarbon chains exposed to the aqueous medium. Because the interactions between the layers are relatively weak, both lipid and protein molecules are able to diffuse rapidly in the plane of the membrane. Membranes can therefore be regarded as fluid structures. The membrane proteins, which act as pumps, receptors and gates, are partially

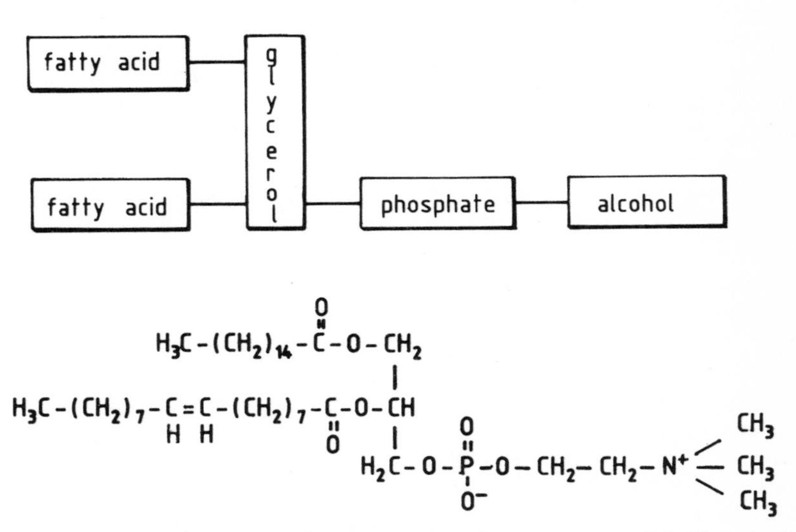

Fig. 4.12 (a) Block diagram of phosphoglyceride structure; (b) Chemical structure of a common phosphoglyceride (1-palmitoyl-2-oleolyphosphatidylcholine).

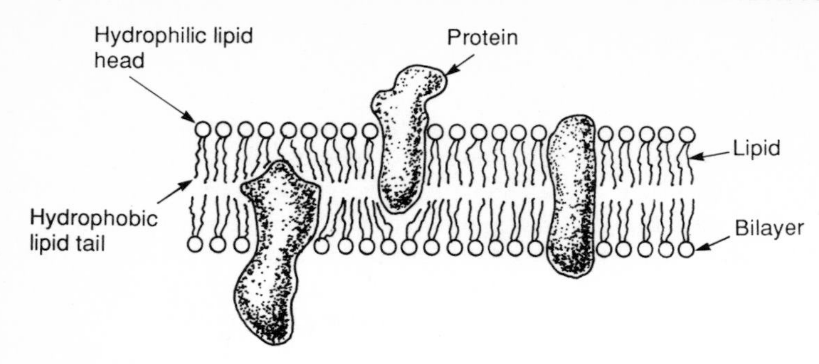

Fig. 4.13 Fluid mosaic model of a lipid membrane structure showing membrane-associated proteins.

embedded in the membrane and partially protruding. Some proteins span the entire thickness of the membrane.

From an initial inspection of lipid structures and their configuration in bimolecular lipid membranes, it appears that the polar phosphate groups alone should elicit a dielectric loss response. However because of the electrically dissimilar nature of the non-polar hydrophobic fatty acid and polar hydrophilic phosphate groups, additional interfacial polarizations have been observed. We have seen in Chapter 3 how Maxwell–Wagner relaxation phenomena occurs wherever there is a build up of charge due to the existence of regions of greatly differing conductivity and permittivity. The two-phase model discussed in Chapter 3 has been extended by Coster and co-workers to take account of the acetyl phase of the lipid as described in Fig. 4.14. In this case the fatty acid tails are regions of low conductivity and dielectric constant and the polar head groups those of high conductivity and dielectric constant with the acetyl regions lying somewhere between the two. Dielectric measurements made on black lipid membranes over the frequency range 0.1 to 100 Hz have indicated the existence of a Maxwell–Wagner dielectric dispersion centred at approximately 10 Hz. Conventional dielectric measurements in the aqueous phase are dominated at these low frequencies by electrode polarization effects and for this reason the four-probe technique, described in Chapter 3 was employed. Analysis of the observed dispersions has provided information regarding the sub-structure of membranes, i.e. the dimensions of these regions, the proportion of solvent present in the membrane, and the effect and extent of the penetration of water and ions into the membrane.

The dielectric properties of a bilayer membrane of the synthetic lipid dioleyl phosphate have been characterized over the frequency range 20 Hz to 3 MHz by employing a hydrophobic filter paper (Millipore) of a given pore size as a substrate on which to form the membrane. A dielectric dispersion was observed for this system centred between 10 kHz and 1 MHz depending on the ionic strength of the suspending KCl solution. This dielectric loss was considered to arise from a

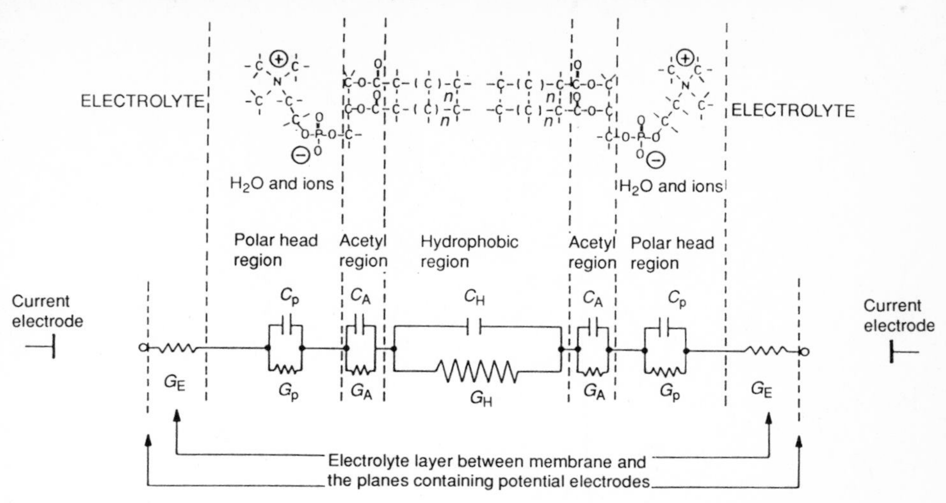

Fig. 4.14 An equivalent circuit for a membrane bilayer/electrolyte system. Each distinct region of the bilayer is represented by an equivalent parallel conductance and capacitance.
From Ashcroft, R.G., Coster, H.G.L. and Smith, J.R. (1981). The molecular organisation of bimolecular lipid membranes, *Biochim. Biophys. Acta* **643**, 191–204.

Maxwell–Wagner type of interfacial polarization process between the membrane and the aqueous ionic phases.

As previously mentioned, lipid head groups are polar in nature and we can therefore expect a dielectric response due to orientational relaxation processes. In some cases the head groups are zwitterionic as shown in Fig. 4.12(b) and the dipole moments are relatively large. Hydrated 1,2-dipalmitoyl-*sn*-glycero-3-phosphocholine (DPPC) for example, has a zwitterionic head group which exhibits orientational motion at the bilayer surface with a characteristic relaxation frequency of approximately 50 MHz. This dielectric property has been employed as a probe for investigating head-group mobility in phospholipid membranes. The dielectric properties of hydrated DPPC bilayers (0.25 g water/g lipid) studied as a function of temperature over the frequency range 1 to 50 MHz indicated sharp discontinuities in both real and imaginary parts of the permittivity at 50 MHz. These are seen, as shown in Fig. 4.15, at a pre-transition (35°C) where a rippled bilayer structure is thought to exist and at the main transition (42°C) reflecting the change from the relatively structurally rigid to the more fluid membrane.

The rotational mobility of the charged head groups of DPPC has also been studied in the presence of cholesterol, which occurs in significant proportions in the membranes of eukaryotes. The relaxation time became shorter and the phase transition less well defined with the addition of cholesterol, these effects being most noticeable at cholesterol concentrations approaching 33%. One of the main factors influencing the mobility of the head groups is the [head group–head group]

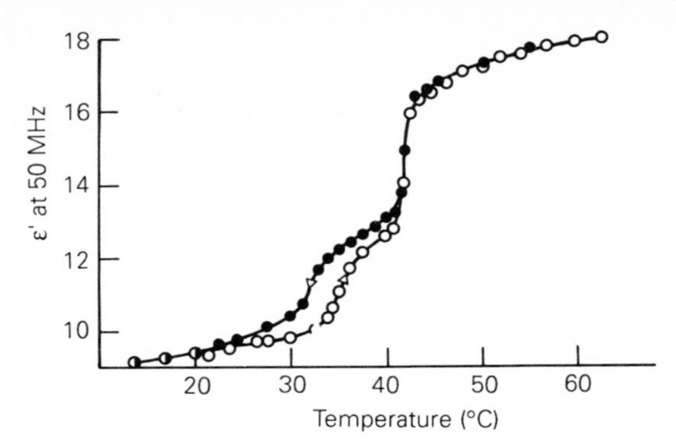

Fig. 4.15 Variation of the permittivity ε' at 50 MHz with temperature for DPPC at 25% water content: (○) increasing temperature, (●) decreasing temperature.
From Shepherd, J.C.W. and Büldt, G. (1978). Zwitterionic dipoles as a dielectric probe for investigating head group mobility in phospholipid membranes, *Biochim. Biophys. Acta* **514**, 83–94.

separation and the dielectric measurements indicate, in agreement with other techniques, that cholesterol has the effect of increasing the head-group separation resulting in a decrease of [head group–head group] electrostatic interaction. This effect alone would result in greater membrane fluidity. However cholesterol also sterically hinders acyl-chain motion and this has the opposite effect of eliciting greater rigidity. We may say, therefore, that cholesterol moderates the fluidity of membranes and dielectric techniques are able to detect that moderating influence.

Liposome and cell suspensions

From lipid bilayer structures we now move on to liposome and cell suspensions. Lipid vesicles, or liposomes, are aqueous spheres enclosed by a lipid bilayer which can be formed by the sonication of a suitable lipid suspended in aqueous medium. Liposomes can be very useful for investigating lipid and membrane properties since their composition can be controlled very precisely in contrast to cells where the membranes tend to be very heterogeneous with varying lipid and protein compositions. The majority of dielectric studies on cell suspensions have been performed on bacterial and yeast cells. In addition to the usual lipid bilayer (plasma membrane), these cells possess cell walls which confer structure and shape to the cell. The two main types of bacteria are classified as Gram positive, having a thick (250 Å) cell wall composed of peptidoglycan, and Gram negative which have a 30 Å thick peptidoglycan cell wall surrounded by an 80 Å membrane of protein, lipid and lipopolysaccharide.

Liposomes and cells are composed from a number of different phases including lipid and aqueous regions which possess very different electrical properties. The

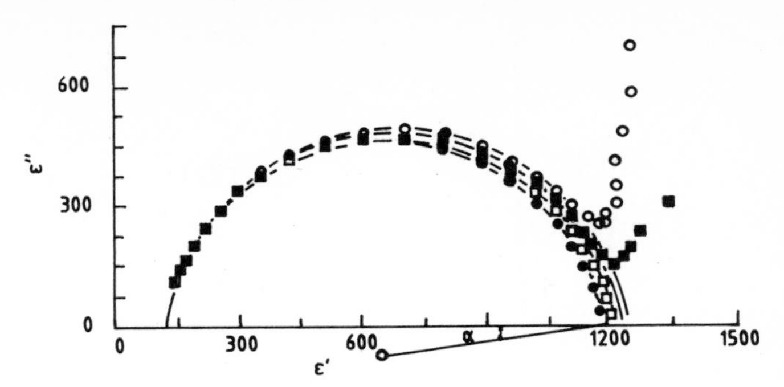

Fig. 4.16 Complex permittivity plot of yeast cells in suspension. D.c. conductivities are as follows: (●) 0.51 mS/cm; (□) 0.505 mS/cm; (■) 0.5 mS/cm; (○) 0.495 mS/cm; α = Cole–Cole parameter indicating the distribution of relaxation times.
From Harris, C.M. and Kell, D.B. (1983). The radio-frequency dielectric properties of yeast cells measured with a rapid, automated, frequency-domain dielectric spectrometer, *Bioelectrochem. Bioenerg.* **11**, 15–28.

accumulation of charged species at the relatively low-conductivity lipid cell membrane can be expected to produce a Maxwell–Wagner type of interfacial relaxation process and this is in fact observed experimentally as the β-dispersion centred in the low megahertz region in the complex permittivity plot of Fig. 4.16 for a suspension of yeast cells. In the past the model most used to simulate this interfacial relaxation was known as the 'single-shell' model (Fig. 4.17a). This consists simply of three phases, that of the membrane, the homogeneous internal compartment and the suspending medium. For spheres surrounded by membranes, assuming the conductance of the membrane is zero, then the following relationships developed by Pauly and Schwan can be applied to describe the β-dispersion,

$$\Delta\varepsilon = \varepsilon_s - \varepsilon_\infty = \frac{9prC_m}{4\varepsilon_o(1+p/2)^2}$$

$$\sigma_l = \sigma_o[1-(3p/2)]$$
$$\sigma_\infty = \sigma_o[1+3p(\sigma_i+\sigma_o)/(\sigma_i+2\sigma_o)]$$
$$\tau = rC_m[(1/\sigma_i)+(1/2\sigma_o)]$$

where r is the cell radius, C_m is the membrane capacitance, p is the volume fraction, σ_l and σ_∞ are the low- and high-frequency conductivities of the suspension and σ_i and σ_o are the conductivities of the cytoplasm and extracellular fluid, respectively. The assumption that the membrane has a negligibly small conductance appears to be valid since most bacterial membranes have conductances of $<10^{-4}$ S/cm^2 and a much higher value (~ 1 S/cm^2) would be required for G_m to have an effect on the experimental results.

Using these equations it has been possible to obtain a value of approximately

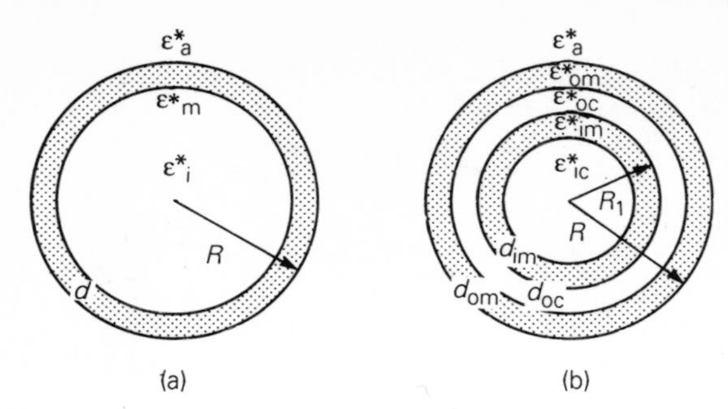

Fig. 4.17 Models used in the dielectric analysis of cell suspensions.
Left: single-shell model. *Right*: double-shell model. ε^* = complex permittivity, R = cell radius, d = thickness.
Subscripts have the following meanings: a, suspending medium; m, membrane; i, internal compartment; om, outer membrane; im, inner membrane; oc, outer compartment; and ic, inner compartment.
From Asami, K. and Irimajiri, A. (1984). Dielectric analysis of mitochondria isolated from rat liver. II. Intact mitochondria as simulated by a double shell model, *Biochim. Biophys. Acta* **778**, 570–578.

1 μF/cm^2 for the membrane capacitance. In addition, the value obtained for the conductivity of the cell interior appears to be consistently lower by a factor of two than would be expected from cytoplasmic ionic-strength considerations. The reason for this is not yet fully understood.

The single-shell model has been employed successfully to predict the dielectric behaviour of liposomes and bacteria which, as shown in the example of Fig. 4.18, do not possess separate nuclei or internal organelles. In these cases the model correctly predicts a dispersion with a single relaxation time and magnitude proportional to the volume fraction. However for eukaryotic cells, an example of which is shown in Fig. 4.19, considerable deviation from this model is apparent and in recent work Asami and Hanai have achieved a better fit to the experimental data by taking into account the presence of the relatively large intracellular nucleus using a 'double-shell' model (Fig. 4.17b).

For cells and vesicles with membranes containing charged phospholipids, a second dielectric dispersion below 1 MHz becomes important. This is known as the 'α-dispersion'. The effect of the presence of charged phospholipids is to attract a population of counter-ions in the solution to the vicinity of the lipid head groups. The α-dispersion is associated both with the movement of these counter-ions tangentially to the vesicle surface, and with the radial polarization of the more diffuse layer of ions surrounding the cell. The theory predicts that the relaxation time should be dependent upon the radius of the particle and the counter-ion mobility, whilst the dielectric increment should be proportional to the radius and to the particle concentration. These dependencies have been observed experimentally. A good agreement between the surface charge density obtained from dielectric measurements and that obtained by microelectrophoresis has also been noted.

A further small dispersion has also been observed for zwitterionic aggregated phospholipids in aqueous and methanolic solutions, centred at a frequency of approximately 100 MHz. This has been interpreted in terms of the diffusional re-orientation of the dipolar phospholipid head groups.

A number of experimental procedures can be followed to help ascertain the processes underlying the observed dielectric dispersions. The equations describing the relaxation frequencies and magnitudes of the α- and β-dispersions have different dependencies upon the vesicle or cell radius. Thus sonication to alter the vesicle radii will produce process-dependent changes in the dielectric characteristics. As detailed in the section on proteins, altering the aqueous solvent viscosity has a considerable effect on the orientational and diffusional processes occurring in that phase. Increasing the aqueous viscosity has the effect of increasing the relaxation time of relaxations due to double-layer ionic motions but should not affect relaxations due to motions in the membrane phase. A procedure which, by contrast, is useful in separating counter-ion from intra-membrane contributions is the addition of cross-linking agents such as glutaraldehyde or dimethyl suberimidate. These have the effect of restricting motions of charged membrane components whilst leaving counter-ion motions unaltered.

Dispersions corresponding to the α- and β-relaxation processes have been described for complete cells, for example for *Paracoccus denitrificans*, *Escherichia coli*, *Micrococcus lysodeikticus*, yeast and erythrocytes. In addition to the α- and β-relaxations, there are other processes which contribute to the dielectric loss. For instance, there will be contributions from intracellular proteins, from organelles and from bound-water and sub-macromolecular species. Dielectric dispersions resulting from these processes are not observed as separate dielectric losses, since they are relatively minor compared with the α- and β-relaxations. However, their contributions may explain the breadth of the β-dispersion observed from bacterial suspensions which appears to indicate an unrealistically large spread of values for C_m, r or σ_i for this loss to be interpreted solely in terms of a Maxwell–Wagner type of process.

The relative magnitudes of the α- and β-dispersions for a particular liposome or cell suspension are dependent on a number of parameters. If the membrane phase contains mainly uncharged lipids, there will be few sites able to attract counter-ions and the α-dispersion will be relatively small. Alternatively, the magnitude of the β-dispersion is dependent on the cell size. If the diameter of the cell is small, the dielectric increment due to the Maxwell–Wagner process may be too small to be detectable especially if a relatively large α-dispersion dominates the dielectric characteristic.

A number of studies have probed the characteristics of the β-dispersion by use of detergents or similar agents to compromise the integrity of the plasma membrane. Treatment of yeast cells with the detergent sodium dodecyl sulphonate (SDS) has been shown to progressively reduce the magnitude of the β-dispersion with increasing concentration of detergent. This is consistent with an increase in the conductance of the membrane and with a decrease in the cell volume.

Dielectric studies by Kell have indicated the existence of another relaxation process, described as the μ-dispersion. This dielectric loss, detected for suspensions of chromatophores from *Rhodopseudomonas capsulata*, was found to be too large,

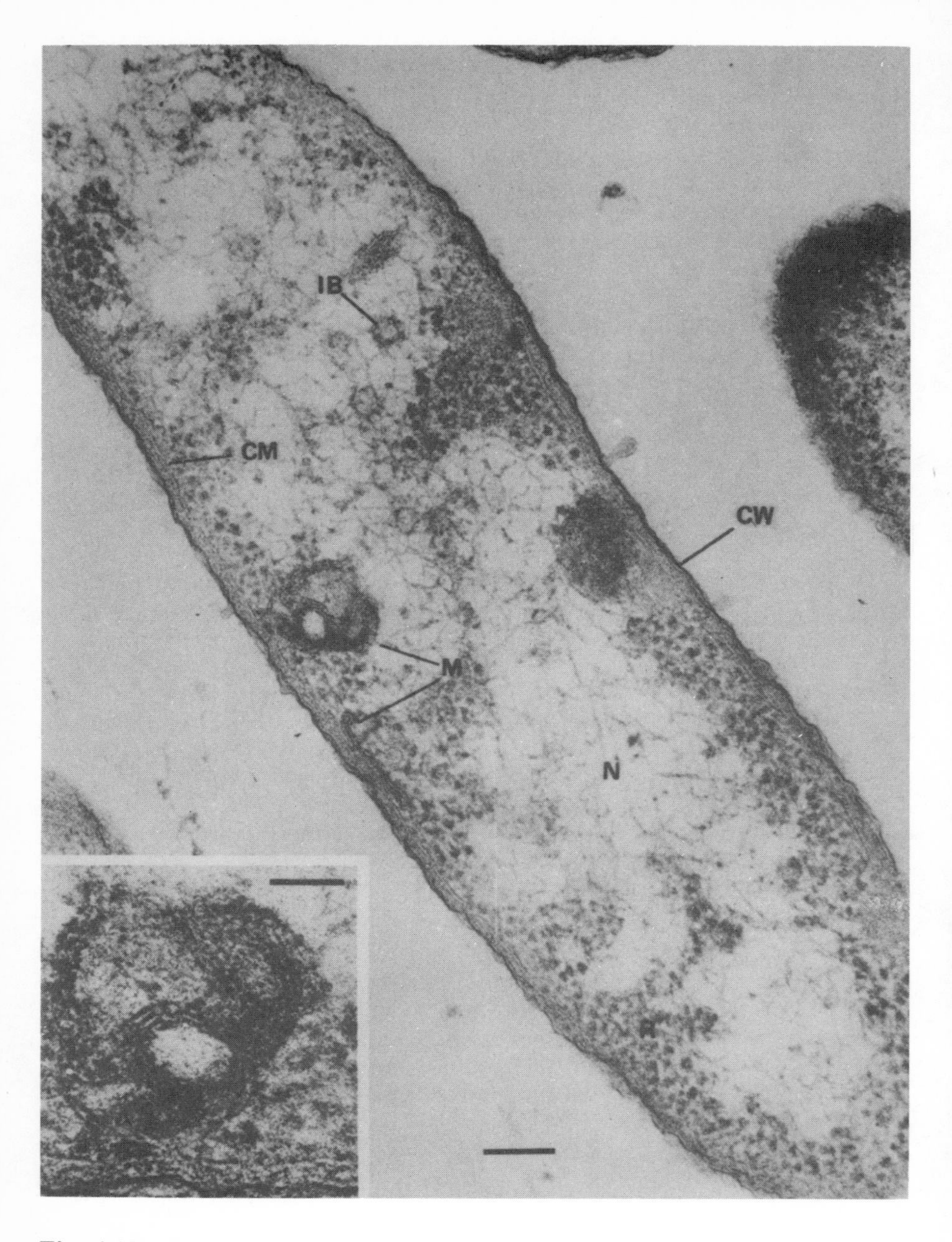

Fig. 4.18 An example of a prokaryotic cell – the bacterium *Pseudomonas aeruginosa*. CM = cytoplasmic membrane; CW = cell wall; IB = inclusion body; M = mesosome; N = nucleoplasm; R = ribosomes. Bar = 100 nm.
From Hoffman, H-P., Geftic, S.G., Heymann, H. and Adair, F. (1973). Mesosomes in *Pseudomonas aeruginosa*, *J. Bacteriol.* **114**(1), 434–438.

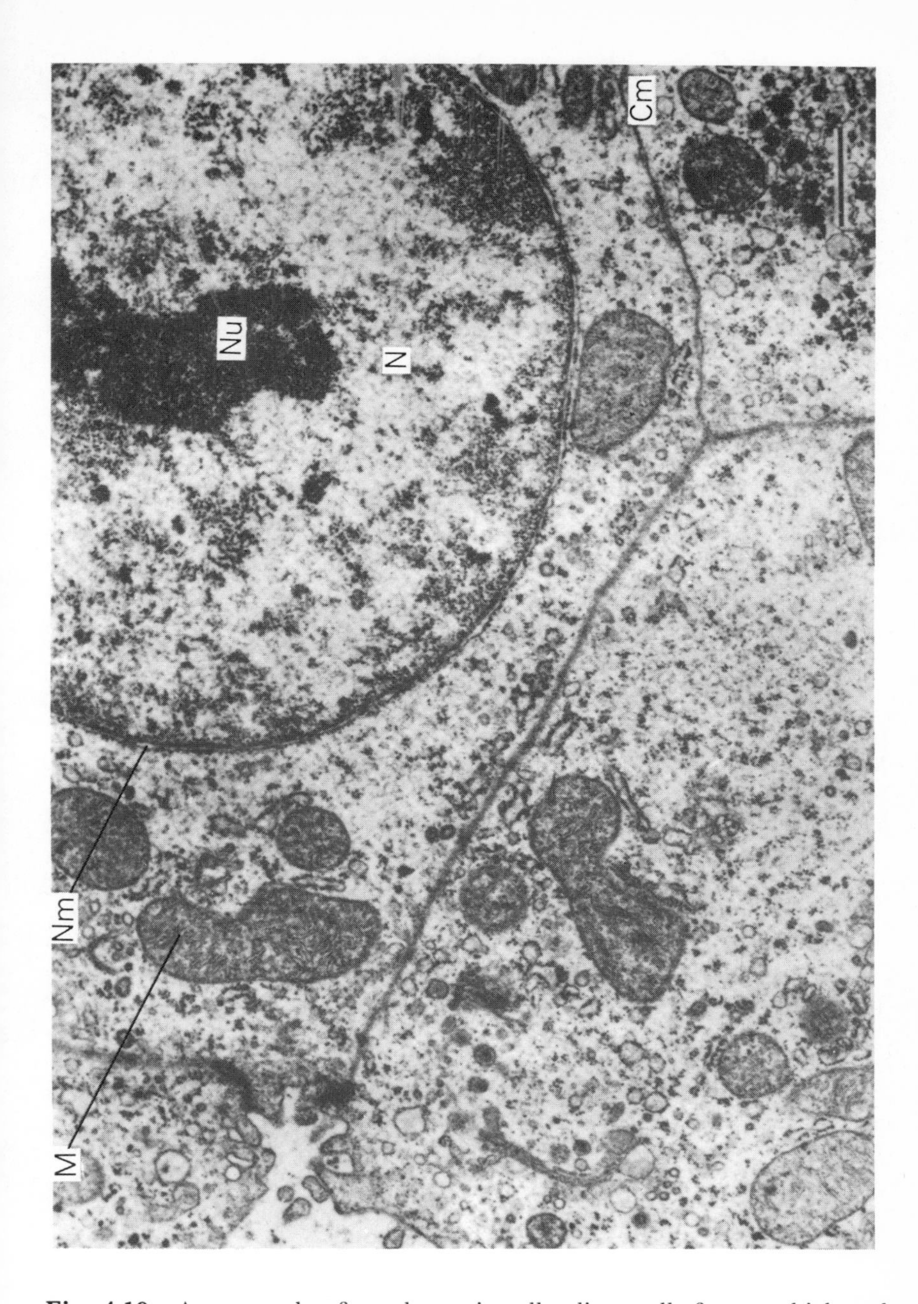

Fig. 4.19 An example of a eukaryotic cell – liver cells from a chick embryo. N = nucleus, Nu = nucleolus, Nm = nuclear membrane, Cm = cytoplasmic membrane, M = mitochondria. Magnification = × 10 500.
From Benzo, C.A. and Nemeth, A.M. (1971). Factors controlling the development of chick embryo liver cells, *J. Cell Biol.* **48**, 236–247.

considering the vesicle size, to be attributed to a Maxwell–Wagner relaxation process. The magnitude of this dispersion was also found to be unaffected by the addition of a detergent which made the cell membrane permeable to ions. This treatment dramatically increases the membrane conductivity and would be expected to significantly reduce the magnitude of a dispersion associated with a Maxwell–Wagner relaxation process. The magnitude of the μ-dispersion was also unaffected by the addition of divalent and trivalent ions which indicated that this was probably not a second counter-ion relaxation process. However, the dispersion was completely destroyed by treatment with the cross-linking agent glutaraldehyde. The main effect of this treatment was considered to be the immobilization of the integral membrane protein component of the cell membranes. The μ-dispersion may therefore be attributed to the diffusional motion of the polar proteins within the plane of the membrane.

Some intracellular organelles such as mitochondria and synatosomes have also been studied using dielectric spectroscopy and have produced similar dielectric characteristics to cells and liposomes. Swollen mitoplasts, for instance, have been analysed using the single-shell model whereas intact mitochondria which have a characteristic double-membrane structure have been simulated using the double-shell model.

Tissue

Tissue is the most complex of all the biological materials discussed. Tissue is composed of cells in an extracellular matrix and as such consists of membranes, macromolecules including structural proteins and DNA, amino acids, electrolytes and water. Obviously with such complex systems the observable dielectric properties are not predominantly determined by relaxation processes attributable to any one molecular species but are dominated, as they were for cell suspensions, by more macroscopic phenomena such as Maxwell–Wagner interfacial polarizations and counter-ion relaxations. There are three principal dielectric dispersions associated with tissue, designated the α-, β- and γ-dispersions as illustrated in Fig. 4.20.

The low-frequency α-dispersion is manifested by a very large increase in the permittivity at low frequencies and is thought to primarily involve the movement of counter-ions tangentially to the charged membrane surface in a way similar to the α-dispersion described for cell and liposome suspensions. It has been suggested that the magnitude of the α-dispersion predicts unusually high fixed-charge densities on the membrane if analysed in terms of a counter-ion relaxation process. An alternative mechanism proposes that this large dielectric loss originates from the build up of charge in the sarcotubular system in the cell cytoplasm, i.e. a Maxwell–Wagner interfacial polarization phenomena within the cell interior. The large extent and complexity of the sarcotubular system could be expected to produce a dispersion at least an order of magnitude larger than the interfacial effects associated with the plasma membrane.

The β-dispersion exhibits a very broad distribution of relaxation times and

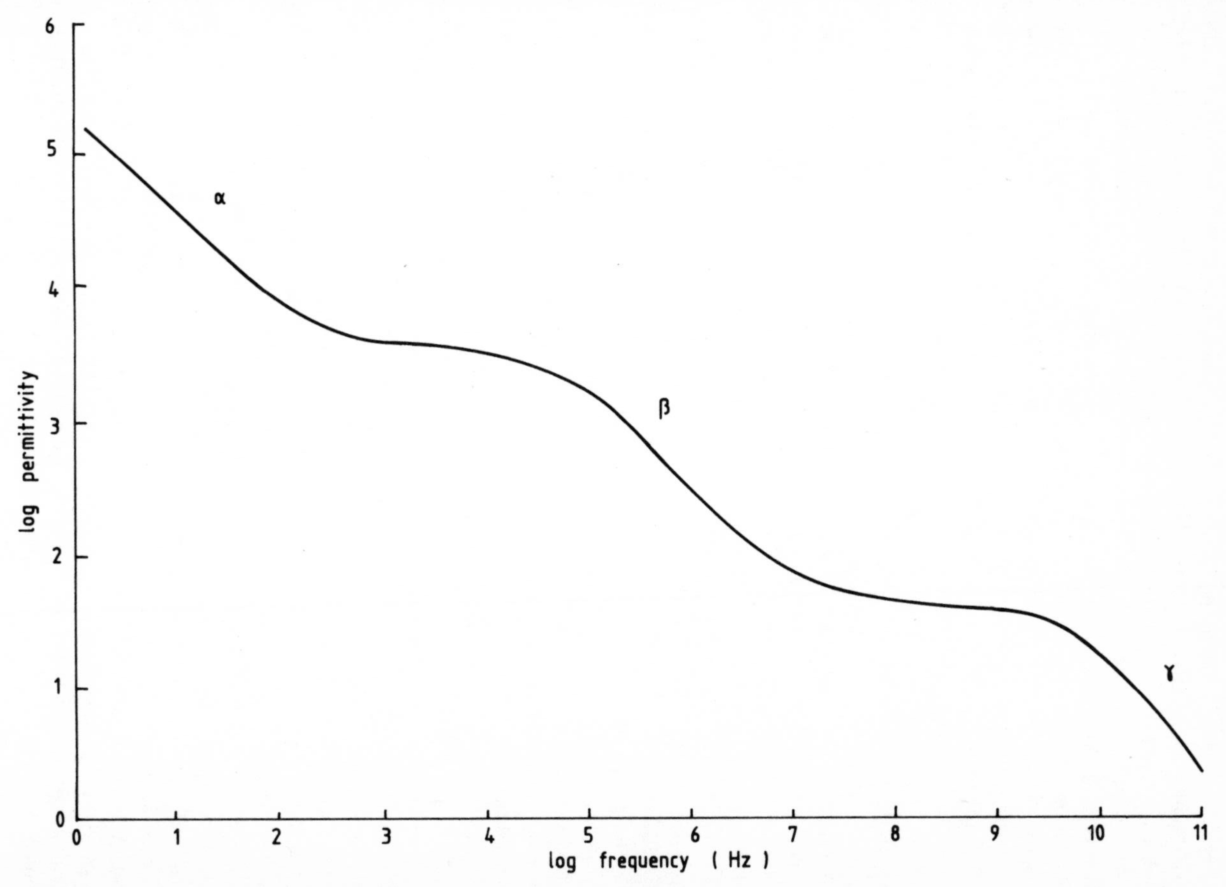

Fig. 4.20 Complex dielectric spectra of a typical tissue sample.

describes at least two relaxation processes. The larger of the two is due to Maxwell–Wagner polarization effects resulting from the build up of charges at low-conductivity boundaries, principally the plasma membranes, within the tissue, whilst a smaller contribution to the loss can be attributed to the orientational relaxation of the soluble protein component as described earlier in this chapter. Some types of tissue do not exhibit distinct α- and β-dispersions, the two loss peaks often merging into a wide, structureless dispersion region.

The high-frequency γ-dispersion centred at about 20 GHz, is associated with the relaxation of water dipoles. Although it has been suggested in some studies that the relaxation time of water in certain tissues is longer than that of normal bulk water by a factor of $\leqslant 1.5$, the majority of dielectric work covering this frequency range has concluded that the observed dispersion is indistinguishable from that of normal bulk water. Recent work has indicated that the high-frequency (>100 MHz) dielectric properties of tissues can be explained in terms of a simple mixture theory with solid matter, bound water and bulk water as components. Differences in the magnitude of the γ-dispersion observed for different tissue types therefore reflect the proportion of water contained within those tissues. This is illustrated in Fig. 4.21 where the dielectric characteristics of muscle (water content 73–77.6%) and fat (water content 5–20%) are described.

Almost all dielectric studies have been performed on excized tissues *in vitro*. However, *in vivo* dielectric properties have also been determined for skin tissue and recently for liver, spleen and kidney. These studies have indicated that the difference in properties between 'live' and freshly-excized tissue is negligibly small. However, at low frequencies, pronounced changes in the dielectric parameters do take place within hours following excision. The magnitude of the α-dispersion decreases significantly with time, the permittivity in the audio frequency range decreasing by almost an order of magnitude within two days of excision. The dielectric parameters defining the β-dispersion, at higher frequencies, change at a much slower rate. It appears from these observations that the sarcotubular system is more sensitive to disruption than the plasma membrane, although the reason for this remains unclear.

There have been a number of studies of the dielectric properties of tumour and homologous normal tissue. It is well known that tumour tissues have significantly higher water contents than the corresponding normal tissue; for instance, normal rat liver has a water content of 71.4% whereas hepatoma has 81.9% water and this is reflected by increased dielectric decrements in dielectric measurements of the γ-dispersion in the gigahertz region.

Many tissue types from a variety of animals have been studied. Although analysis of the dielectric properties in terms of the properties of individual molecular components is not usually possible, information regarding permittivities and conductivities of the various tissue types has proved invaluable in medical research concerned with, for instance, the radio frequency hyperthermal treatment of cancer.

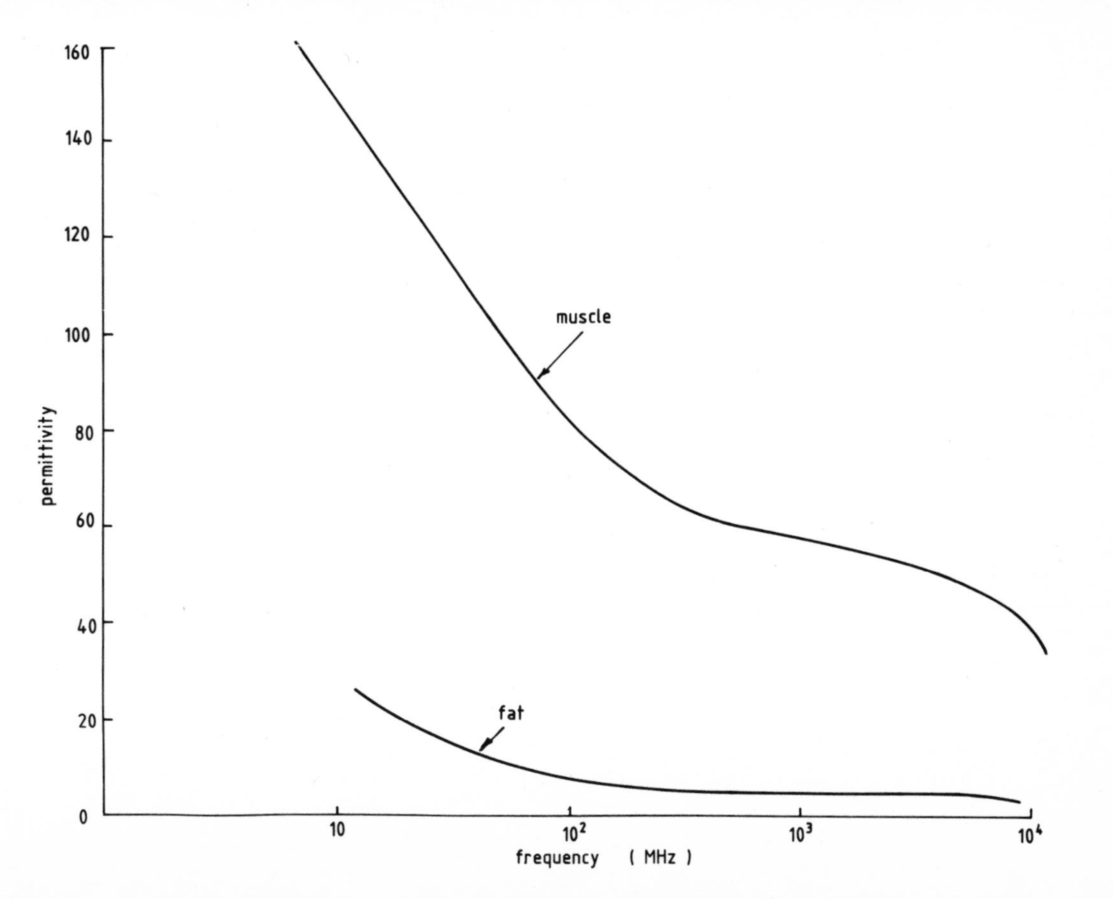

Fig. 4.21 Dielectric permittivity characteristic at microwave frequencies for muscle and fat tissue.

Selected reading

Asami, K., Takahashi, Y. and Takashima, S. (1989). Dielectric properties of mouse lymphocytes and erythrocytes, *Biochim. Biophys. Acta* **1010**, 49–55.

Foster, K.R. and Schwan, H.P. (1986). Dielectric properties of tissues, in *C.R.C. Handbook of Biological Effects of Electromagnetic Fields*, Eds Polk, C. and Postow, R., Florida, CRC Press, pp. 28–96.

Foster, K.R., Schepps, J.L. and Epstein, B.R. (1982). Microwave dielectric studies on proteins, tissues and heterogeneous suspensions, *Bioelectromagnetics* **3**, 29–43.

Grant, E.H., Sheppard, R.J. and South, G.P. (1978). *Dielectric Behaviour of Biological Molecules in Solution*. Oxford, Clarendon Press.

Harris, C.M. and Kell, D.B. (1985). On the dielectrically observable consequences of the diffusional motions of lipids and proteins in membranes I and II, *Euro. Biophys. J.* **12**, 181–197 *ibid.*, **13**, 11–24.

Kell, D.B. and Harris, C.M. (1985). Dielectric spectroscopy and membrane organisation, *J. Bioelectricity* **4**(2), 317–348.

Pethig, R. (1979). *Dielectric and Electronic Properties of Biological Materials*. John Wiley.

South, G.P. and Grant, E.H. (1972). Dielectric Dispersion and Dipole Moment of Myoglobin in Water, *Proc. R. Soc. Lond. A* **328**, 371–387.

Stoy, R.D., Foster, K.R. and Schwan, H.P. (1982). Dielectric properties of mammalian tissue from 0.1 to 100 MHz: a summary of recent data, *Phys. Med. Biol.* **27**, 501–513.

Takashima, S., Gabriel, C., Sheppard, R.J. and Grant, E.H. (1984). Dielectric behaviour of DNA solutions at radio and microwave frequencies, *Biophysics J.* **46**, 29–34.

Chapter 5

Current Dielectric Studies of Biological Molecules and Systems

Hydration studies

DIELECTRIC PROPERTIES OF BIOLOGICAL WATER

In the past, water was regarded as little more than a suspending medium for the active biological molecules with little or no significant contribution to biological processes. It has only recently become recognized that water may play an important and fundamental role in determining the physical and chemical properties of proteins which are ultimately responsible for their unique biological activity. The way in which water interacts with a protein is determined by its polar character and the drive to achieve the lowest energy state by maximizing the number of hydrogen bonds with the protein and with other water molecules. This hydrogen-bonding ability makes water a material with some remarkable properties. It exhibits a considerably higher specific heat, latent heat of vaporization and surface tension than similar molecules such as H_2S and melting and boiling points some 100°C lower than expected from molecular size considerations. As a consequence of its polar character, water also has a high dielectric constant and has the ability to weaken other electrostatic interactions including those between polar groups in the protein structure. Despite many decades of investigation, the structure of liquid water remains the subject of much controversy. Models range from highly coordinated hydrogen bonding schemes where virtually all possible hydrogen bonds are made to flickering cluster models in which pentameric or tetrameric groups of hydrogen-bonded water molecules fluctuate very rapidly through a number of different hydrogen bonding configurations.

Water molecules have a dipole moment of 1.84 Debye (6.14×10^{-30}C.m) and in the normal bulk state exhibit a room temperature (25 °C) dielectric dispersion centred at 17 GHz. A temperature-dependent relaxation activation energy of ~4.5 kcal/mole has been measured and this is consistent with a mechanism involving molecular rotation. Since the relaxation time is dependent upon the activation energy required for rotation of the molecule, any change in the nature of the intermolecular interactions experienced by the water will affect the rotational correlation times. For proteins in aqueous protein solution, water molecules close to the protein exhibit increased hydrogen-bonding interactions with the protein and with other water molecules, and for this reason such water molecules possess longer characteristic relaxation times than that of normal bulk water.

Water molecules are bound to biomacromolecules by hydrogen bonding to polar, hydrophilic sites. In the case of proteins these sites are located on peptide units and acidic and basic side groups. Water bound to biomacromolecules is rotationally hindered and has been shown to exhibit a substantially increased relaxation time compared with that of water in the normal bulk state. In addition to primary-bound water involved in single and multiple hydrogen bonding with the protein, a number of studies have shown that the influence of the protein on the water structure extends a considerable distance from the protein surface. Layers of water molecules which are perturbed by the macromolecule and therefore called 'structured water' can be considered to surround the protein. The nature and extent of the protein-bound and structured water and the elucidation of group hydration in protein structures has been the subject of many studies employing nuclear magnetic resonance, infra-red and raman spectroscopy, calorimetry and electron spin resonance as well as dielectric techniques.

ASPECTS OF PROTEIN HYDRATION

In the case of proteins, the ionizable groups (COO^- and NH_3^+) appear to be the primary hydration sites. The existence of these soluble, ionizable residues allows the initial condensation of water molecules onto the protein and the hydration and ionization of these are the primary processes in protein hydration. The COO^- groups of aspartic (Asp) and glutamic (Glu) acids are able to bind on average two water molecules each. Some Glu and Asp groups which make fewer hydrogen bonds with the protein are able to bind three or even four water molecules.

Water is also bound to the polar (non-ionizable) side chains and to the polar peptide units of the protein backbone and this process is competitive. The contribution of the peptide groups to protein hydration has proved difficult to assess but gravimetric studies on polypeptides with apolar side groups such as polyglycine seem to indicate that where the backbone is accessible to the aqueous environment, at least one water molecule is bound by each peptide unit. Most polar, non-ionizable amino acid residues tend to bind an average of one water molecule per side chain. Solvent accessibility may be a determining factor in the extent to which particular side groups are hydrated. By virtue of their lengths, lysine, arginine and glutamic acid are most likely to protrude into the solvent. By contrast, histidine and aspartic acid often have low solvent accessibilities as a result of a bulky aromatic ring and short side chain, respectively.

The hydrophilic groups are not randomly arranged in the protein structure. This is due to the fact that there is a favourable decrease in the protein's free energy when non-polar groups are withdrawn from contact with an aqueous environment to the hydrophobic protein interior and it is then energetically unfavourable for polar groups to be buried in a dehydrated state in the non-polar interior of the protein. So non-polar groups tend to lie in the protein interior and polar ones near to, or on, the protein surface. In general, therefore, protein hydration can be said to take place predominantly on the protein surface.

There are notable exceptions to this rule. Even small proteins can contain buried water molecules, which may occupy internal cavities. Lysozyme, for example, incorporates four internal water molecules and cytochrome *c* has three. These internal water molecules nearly always electrostatically compensate internal polar or charged groups by forming hydrogen bonds with them, thus stabilizing the local protein structure. A number of proteins contain extensive internal water networks, for instance eight water molecules form an internal cluster in actinidin, as shown in Fig. 5.1(c). Also where two or more internal polar groups are unable to make direct hydrogen bonding contact, water molecules often form a bridge between them (Fig. 5.1(b)). Although this water is situated in the interior of the protein, solvent exchange experiments have shown that it is still accessible to the bulk water environment, although the exchange rates are much reduced compared with those of surface-bound water. In addition to buried internal water, there are other water molecules which, while being tightly bound to the protein, are also accessible to the bulk aqueous environment. These are often located in surface pockets and crevices, as shown for actinidin in Fig. 5.1(a).

Although there is a tendency for hydrophobic groups to be internal and hydrophilic groups to be on the protein surface, this is a generalization. Hydrophobic groups are found on the surface of the great majority of proteins. In fact, many proteins possess considerable areas of surface hydrophobicity since there is also a tendency for clusters to form, i.e. for non-polar groups to be preferentially surrounded by other non-polar groups and likewise for polar groups. In these areas of surface hydrophobicity, hydration takes the form of the spreading of water clusters from the ionized and polar groups to cover the hydrophobic regions. It seems most likely that water bridges form over the non-polar region.

Water plays a crucial role in the conversion of the unfolded nascent polypeptide chain into the active, native protein. Primarily the native structure is under the direction of the amino acid sequence itself. A variety of intramolecular interactions influence the final protein conformation – electrostatic and torsional interactions, dispersion forces and most importantly solvent interaction. Polar water molecules compete with possible intramolecular interactions between polar groups of the protein and also take part in hydrophobic interactions with non-polar amino acid side groups, making the protein interior a more energetically favourable environment for these species. Although the peptide backbone is composed of polar amide and carbonyl groups, interaction with the aqueous environment is sometimes not the most energetically favourable option. Hydrogen bonding between the amide nitrogen and carbonyl oxygen in a helical or sheet structure provides for electrostatic compensation and is relatively strong in a non-aqueous environment. Also, as we have seen, the interaction of water molecules with the

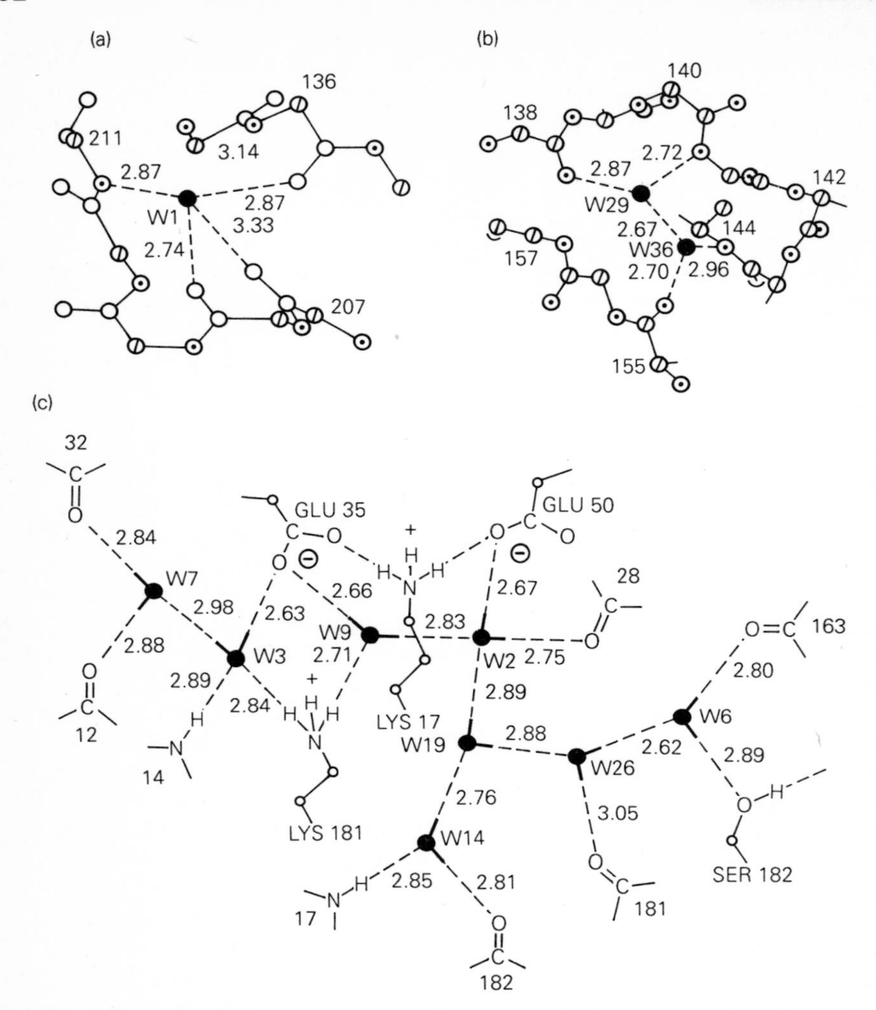

Fig. 5.1 Typical water sites from the structure of actinidin. (a) Discrete internal site where the water molecule is within hydrogen bond distance of three C=O (acceptors) and two N–H (donors). (b) Pair of water molecules filling a surface crevice between two sections of polypeptide chain. (c) Cluster of eight water molecules forming a network of hydrogen bonds in the interior of the actinidin molecule.

From Baker, E.N. and Hubbard, R.E. (1984). Hydrogen bonding in globular proteins, *Prog. Biophys. Molec. Biol.* **44**, 97–179.

peptide components, particularly the main chain amides, is, for geometric reasons, relatively weak. In homopolypeptides, where the side chains do not impose structural irregularity, this results in helical conformations where there is maximization of the hydrogen bonding between the peptide components.

DIELECTRIC STUDIES OF PROTEIN HYDRATION

Hydration phenomena have been studied using dielectric measurements on both aqueous solutions and on biomolecules in the hydrated solid state. Solution studies have proved most useful in assessing the total water perturbed by the biomolecule in its native environment. On the other hand hydrated solid-state studies have been sufficiently sensitive to provide detailed information regarding the nature and extent of bound water populations but have the disadvantage of having been carried out in non-biological conditions.

PROTEIN SOLUTION STUDIES

The dielectric characteristics of proteins in solution have been extensively studied over the last forty years. Measurements over the frequency range 50 kHz to 20 GHz have indicated the presence of three dielectric dispersions. These are the so-called *β-dispersion* (centred in the low megahertz region), the *δ-dispersion* (which is a relatively small dielectric loss extending over a wide frequency range centred at approximately 100 MHz) and the *γ-dispersion* (located at 17 GHz) (Fig. 4.9; see p. 71). As discussed in Chapter 4, the β-dispersion can be attributed to the rotation of protein molecules in the aqueous medium and the γ-dispersion to the relaxation of bulk water. A number of mechanisms have been proposed for the δ-dispersion but it is now widely believed that the relaxation of protein-bound water is one of the principal processes involved. Other polarization mechanisms which may also contribute include side-chain motions and proton fluctuations on the protein surface.

A number of dielectric studies including a detailed analysis of dielectric data on myoglobin in solution over the frequency range 100 kHz to 18 GHz and for temperatures between 9 and 45 °C has indicated that the δ-dispersion is the combination of two separate dielectric loss peaks with relaxation frequencies of 20 MHz (δ_1) and 3 GHz (δ_2). Analysis of these two dispersions along with the β- and γ-losses can produce valuable information about the hydration properties of the protein.

From X-ray data it is known that the myoglobin molecule is disc-shaped with an axial ratio of 2:1 which should exhibit a single rotational relaxation time and dielectric properties virtually indistinguishable from those of a spherical-shaped molecule. Using this information the β-dispersion data can be applied to the Debye equation

$$\tau = 4\pi\eta a^3/kT$$

which relates the molecular orientational relaxation time τ to the solvent viscosity, η, and the radius of the spherical solute molecule, a. From this equation, a value of the radius of the myoglobin molecule has been calculated to be 2.12 nm which is equivalent to a molecular volume of 41 nm^3. This is considerably larger than the value of 17 nm^3 obtained from the X-ray analysis of anhydrous myoglobin. The difference in volumes is attributed to protein-bound and protein-structured water which is assumed to be considerably more viscous than normal liquid water and hence is constrained to rotate along with the protein as a single entity in the megahertz frequency range. The β-dispersion can, therefore, be said to describe the rotational relaxation of the hydrated protein molecule in solution. The value of the protein radius derived from the Debye equation is only approximate since this relation makes the assumption that the solute molecule is spherical. If the asymmetry of the myoglobin molecule is taken into account then the additional volume is consistent with there being a hydration shell of 0.35 nm or an average of approximately two water molecules' thickness surrounding each myoglobin molecule. This is equivalent in terms of mass to a hydration of 0.6 g water/g protein.

If we now turn to the δ-dispersions centred at 20 MHz and 3 GHz. It is generally agreed that the higher frequency loss, i.e. the δ_2-dispersion, can be attributed to the relaxation of protein-bound water molecules which exhibit independent rotational motion in this frequency region. By applying a suitable mixture model the magnitude of the decrement for this dispersion is found to correspond to approximately 0.3 g water/g protein relaxing with this particular characteristic relaxation time. The origin of the δ_1-dispersion centred at 20 MHz is the subject of some controversy. This dispersion exhibits a very similar dielectric decrement to that of the δ_2-dispersion and would, if analysed in terms of protein-bound water, also produce a hydration value close to 0.3 g/g. Taken together, the δ-dispersions might, therefore, be considered to result from the existence of two components of protein-bound water with their different relaxation times perhaps reflecting different binding energies to the protein. The total amount of protein-bound water would then also be consistent with that predicted by analysis of the β-dispersion (0.6 g water/g protein). An alternative scheme proposed is that the δ_1-dispersion is the result of polar side-chain rotation. The water of hydration predicted by the β-dispersion but not accounted for in this scheme might be irrotationally bound to the protein and therefore unable to exhibit independent relaxation phenomena. If we consider the nature of the bonding involved between water and proteins, it seems reasonable to assume that multiple hydrogen bonded water might be in such a category.

Similar analysis has been used to investigate the hydration components of a number of proteins in solution such as a haemoglobin, ribonuclease and the peptide insulin. Although important hydration data has been obtained, some of the problems inherent in such analysis should be mentioned. The main weakness of the approach is that the hydration values obtained are, to a large extent, dependent on the mixture model used. The analysis is also dependent upon the shape of the protein which may not be known accurately. In addition, the model usually used to describe protein hydration in solution studies, that of a shell of water surrounding the protein, is perhaps rather simple when one considers the discrete nature of water binding and the complexity of the protein surface.

PROTEIN POWDER STUDIES

Although valuable information regarding protein-bound water can be obtained from dielectric studies on solutions, studies involving the stepwise hydration of proteins in the solid state have proved more sensitive and less problematic. These studies have been especially useful for the investigation of relaxation processes which produce small dielectric losses such as those due to protein-bound water and side-chain rotations since the two larger dispersions due to rotation of the protein molecule (β) and bulk water (γ) which tend to dominate in solution, are absent in solid state work. A further advantage is that the ionic steady-state conductance component and large capacitance effects associated with ionic double layers at the electrodes which often cause problems in solution experiments, are much smaller for hydrated powder samples. This allows very small dielectric dispersions to be measured with great accuracy down to relatively low frequencies (<1 Hz) whereas experiments involving biomolecules in solution usually have a low frequency limit of $\geqslant 1$ kHz depending on the ionic strength of the sample solution.

The main objection to studies on powder samples has been that this dehydrated form does not represent the natural state of the biomacromolecule and relatively large conformational differences may exist between the protein in its native solution state compared to the dry-powder form. Details of the effect of hydration on the three-dimensional structure of biomacromolecules are scarce and often inconclusive. Most studies on this subject have been conducted on the enzyme lysozyme. The infra-red and Raman experimental data available to date seems to indicate that there are some conformational changes over the hydration range up to 0.1 g water/g lysozyme but that from 0.1 g/g to dilute solution, large structural changes do not occur. It must be pointed out however that lysozyme is one of the more stable proteins and as such may not be typical of biomacromolecules in general.

A number of studies of protein hydration have been performed over the last twenty years or so by taking dielectric measurements over the frequency range 1 MHz to 20 GHz for incremental increases in protein hydration. Although considerable differences remain between studies as to the precise relaxation times of observed dispersions and how these should be interpreted, it is generally agreed that two relaxation processes occur in this frequency range. The higher frequency dispersion, observed in the low gigahertz region, is usually attributed to protein-bound water whilst the lower frequency relaxation, centred around 10 MHz, has been interpreted alternatively in terms of side-chain relaxations, further rotationally hindered water or ion/proton fluctuations.

One important observation made in these studies is that water molecules bound initially to the protein, i.e. those associated with the protein at hydrations up to say 5% water by weight, do not make any contribution to the polarizability of the protein. This implies that these water molecules are unable to respond to the electric field and are generally thought to be irrotationally bound to the protein structure.

In a recent study on the enzyme chymotrypsin, the dielectric properties were measured for stepwise increases in hydration. Two separate dielectric dispersions

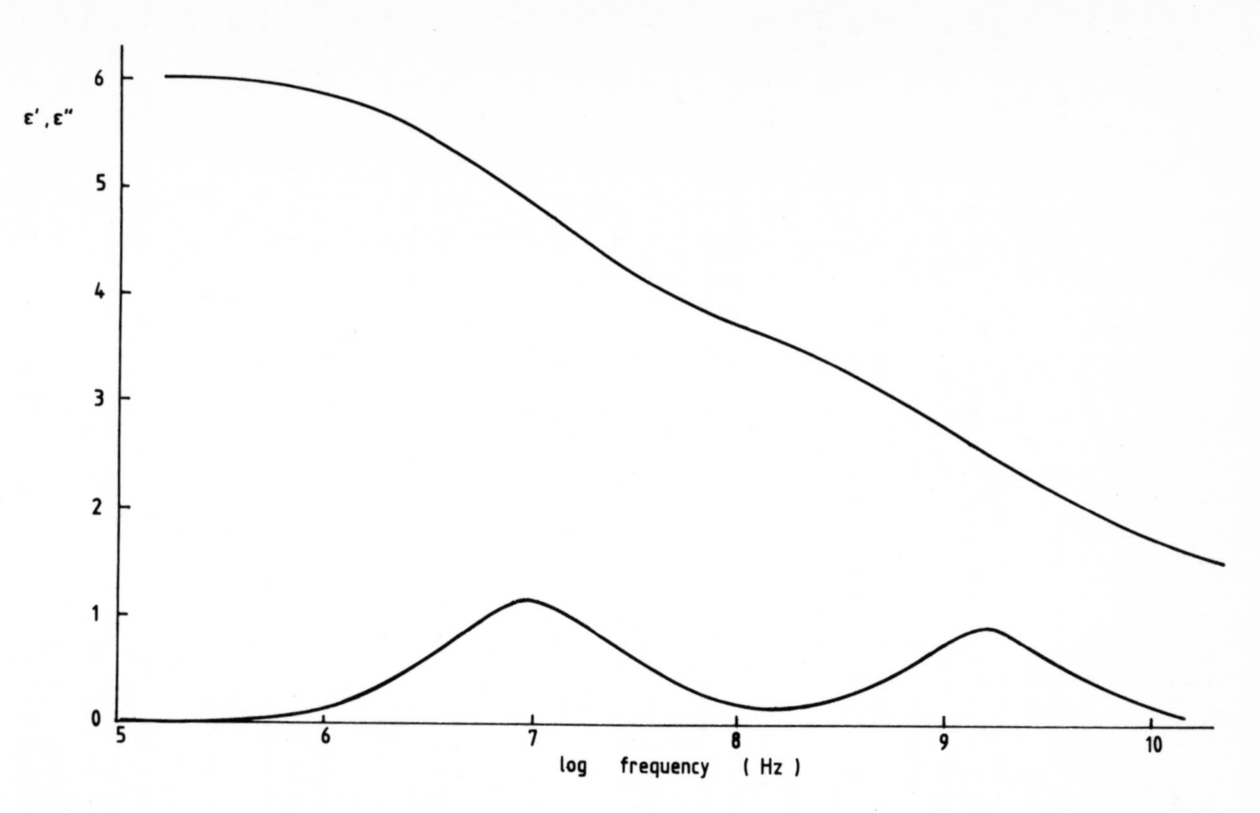

Fig. 5.2 Complex permittivity spectra of hydrated chymotrypsin (370 water molecules per chymotrypsin).

were observed centred at 10 MHz and approximately 4 GHz, as shown in Fig. 5.2 for a hydration value of 370 water molecules per chymotrypsin. Using parallel dielectric and gravimetric measurements and correcting for the local electric field using the Onsager correction, it was possible to monitor the effective dipole moment of the bound water molecules over the hydration range employed using the relation

$$\frac{\mathcal{N}\bar{\mu}^2}{9\varepsilon_o kT} = \frac{(\varepsilon_s - \varepsilon_\infty)(2\varepsilon_s + \varepsilon_\infty)}{\varepsilon_s(\varepsilon_\infty + 2)^2} = \mathrm{F}(\varepsilon) \tag{5.1}$$

where $\mathcal{N}$ is the number density of water molecules bound to the protein, $\bar{\mu}$ is the mean effective dipole moment and ε_s and ε_∞ are the permittivity values at the low and high frequencies, respectively, and which define the boundaries of the dispersion. The increase in the value of the right-hand side of equation (5.1), denoted by F(ε), with increasing hydration was considered to reflect an increase in the polarizability of the protein sample, either due to increasing content of dipolar water molecules, to changes in the vibrational freedom of the protein structure, or to the onset of proton or ion fluctuations on the protein surface.

The variation of the value of the dielectric parameter F(ε) for the high-frequency dielectric loss as a function of the amount of protein-bound water observed in this work is shown in Fig. 5.3. The associated dispersion was considered to arise from the rotation of bound water and the dashed line represents the gradient expected if the protein-bound water molecules behaved as normal bulk water with $\bar{\mu} = 6.14 \times 10^{-30}$C.m.

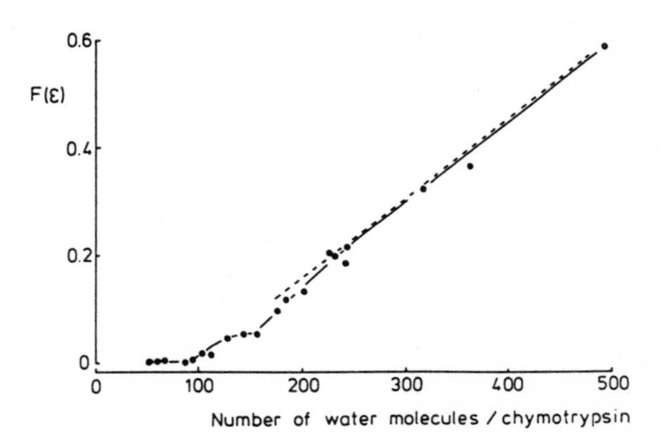

Fig. 5.3 Variation of the dielectric parameter F(ε) of equation (5.1) for the dielectric loss above 100 MHz as a function of the number of water molecules per chymotrypsin molecule. The broken line represents the characteristic expected if the protein-bound water were to possess an effective dipole moment identical to that of normal bulk water.

From Bone, S. (1987). Time-domain reflectometry studies of water binding and structural flexibility in chymotrypsin, *Biochim. Biophys. Acta* **916**, 128–134.

As can be seen in Fig. 5.3, for values of water content up to 90 water molecules per chymotrypsin, the magnitude of F(ε) was found to be relatively insensitive to hydration. This suggests that each of these water molecules is tightly bound to the protein in such a way as to prevent it from rotating. A possible physical model is that each of the water molecules in this group is bound by two or more hydrogen bonds. These water molecules, together with others which remained strongly bound even after dehydration, can be envisaged as being tightly incorporated into the vibrating protein structure with relaxation times equivalent to that of the protein.

At a hydration level corresponding to ~90 water molecules per protein, a break in the characteristic described in Fig. 5.3 was observed. This break occurred at a hydration value very close to the monolayer value obtained from the hydration isotherms, corresponding to the level at which the primary sorption sites on the protein are filled. At hydrations immediately >90 water molecules per chymotrypsin, discontinuity was evident prior to a steep rise in permittivity with hydration >160 water molecules. Similar discontinuities have been observed for lysozyme in the hydration dependence of the absorbance of the carboxylate band maximum and in the frequency of the highest intensity maximum of the OD stretching band of absorbed D_2O. This was assumed to be the result of a rearrangement of arrays of water molecules bound to the soluble, ionizable residues to form clusters of water molecules possessing a more ordered, stable surface structure.

In one proposed hydration scheme, the binding of water to proteins at low hydrations has been discussed in terms of nucleation of water droplets. In this model, the existence of soluble ionizable residues allows the initial condensation of water molecules onto the protein and the hydration and ionization of these are the primary processes in protein hydration. As the number of water molecules bound to the proteins' ionizable groups increases, a point is reached at which a transition occurs in the water structure. This involves the rearrangement of disordered arrays of water dipoles bound to the ionizable groups to form clusters possessing a more ordered, stable surface structure largely determined by surface tension. Transitions of this kind over the protein surface may determine the hydration-dependent dielectric characteristics over the hydration range 90–160 water molecules per chymotrypsin.

The distinct break in the hydration characteristic at ~160 water molecules was followed by a steep rise in F(ε) with increasing hydration. The water bound in this hydration region appears to exhibit an effective dipole moment comparable with that of normal bulk water. This would seem to imply that these water molecules are predominantly involved in relatively weaker interactions with the protein. These water molecules are considered to be singly hydrogen-bonded to the protein so that their rotational motion is relatively unhindered. Over this hydration region, it is likely that the growth of water clusters around the ionizable groups leads to increasing interaction with neighbouring non-ionizable side chains and the protein backbone.

The categorization of protein-bound water from steps in the hydration-dependent properties requires a certain degree of prudence since it is likely that the behaviour of water associated with the partially hydrated protein is altered to some extent for the fully-hydrated state. Good correlation has, however, been observed between the number of water molecules irrotationally bound to lysozyme in dielectric studies on the partially hydrated protein and the number forming multiple hydrogen bonds with lysozyme in X-ray studies on protein crystals, where the amount of water is relatively high.

A second dielectric dispersion was also observed in this work on hydrated chymotrypsin centred at approximately 10 MHz. This dielectric loss was interpreted in terms of the rotation of polar components of the protein structure although the movement of ions or protons on the protein surface could not be discounted as a possible mechanism. In this case the observed increase in the polarizability with increasing hydration, described in Fig. 5.4, would reflect a hydration-induced increase in the flexibility of the protein structure. As discussed later in this chapter, the effect of water on the dynamics of protein structures may have important implications in biological activity.

Another protein whose hydration characteristics have been studied in a similar way is the enzyme lysozyme. In this case there is a considerable amount of hydration data from other techniques with which to compare results obtained from dielectric measurements. For a stepwise increase in lysozyme-water content, the following hydration scheme was observed. For water bound to the protein up to a hydration corresponding to 0.04 g water/g protein, no significant dielectric loss is apparent. These water molecules are bound to the primary sorption sites of the protein and are irrotationally bound to the protein structure by two or more hydrogen bonds. For the hydration region corresponding to between 0.04 and 0.13 g/g water associated with each lysozyme molecule, a steady increase in the

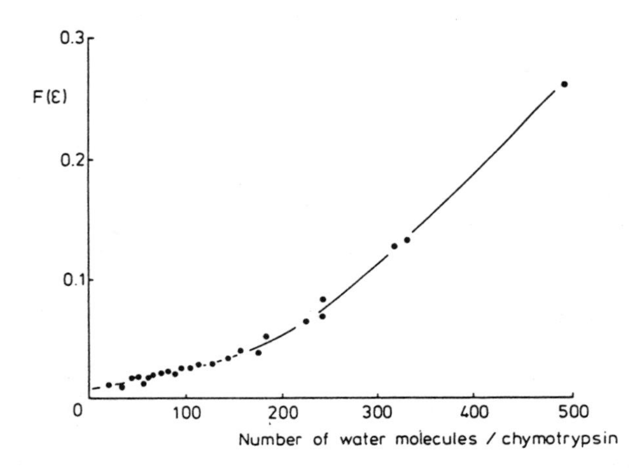

Fig. 5.4 Variation of the dielectric parameter F(ε) of equation (5.1) as a function of hydration of chymotrypsin for the dielectric dispersion centred at 10 MHz.

From Bone, S. (1987). *Biochim. Biophys. Acta* **916**, 128–134.

polarizability of the enzyme is observed. The water molecules in this hydration region are considered to be bound to the protein by single hydrogen bonds in such a way as to be free to move in a relatively unhindered manner. Analysis of the rate of increase in polarizability with hydration indicates that these water molecules have an effective dipole moment only 6% less than the dipole moment of normal bulk water molecules.

These results are in good agreement with the conclusions of recent X-ray work on tortoise egg-white lysozyme which indicated the existence of 31 water molecules (0.04 g/g) making two or more hydrogen bonds and a further 80 (up to 0.14 g/g) making single hydrogen bonds to each lysozyme molecule.

Additional information as to the state of bound water in terms of its translational and rotational properties compared with those of normal bulk water have come from X-ray crystallographic studies. In protein crystals, approximately half the total volume is occupied by water, the protein molecules being surrounded largely by water. Although there can be difficulties resolving the water molecules even in high-resolution studies, a number of well-resolved solvent peaks have been reported for several proteins. Water molecules located inside the protein have been those most easily defined, being hydrogen-bonded to a number of buried polar side groups or to other water molecules and therefore relatively rotationally immobile. Solvent peaks allocated to water molecules bound by two or more hydrogen bonds have also been reported located in the active sites of a number of enzymes, and also on the surface of a number of different proteins where up to 82 water molecules per protein have been detected. Solvent peaks which might be attributed to water molecules bound by single hydrogen bonds to the protein are much less well defined and it appears that water molecules appear 'fixed', as far as X-ray studies are concerned, only if they are anchored by multiple hydrogen bonding to the protein. The majority of the solvent in the protein crystal appears to be relatively unperturbed by the protein, indicated by the absence of well-resolved electron density peaks over most regions not occupied by the protein molecule.

The scheme described here is not the complete picture since the dynamic nature of the molecules involved has not been discussed. The water molecules involved in binding to the protein and in the structured water component are in dynamic equilibrium and are constantly exchanging with the bulk water environment. From the translational information provided by nuclear magnetic resonance work, it is apparent that all protein-bound water, including internal water, is able to exchange with the aqueous phase. However, it is likely that the translational lifetimes of the bound-water molecules differ considerably from those in the bulk state.

Nuclear magnetic resonance, X-ray crystallography and dielectric studies have indicated that there is a relatively large population of water bound to the protein which exhibits very similar properties to normal bulk water. In spite of this, the freezing properties of this water are not normal in that the so-called non-freezing fraction of protein-bound water exhibits a considerable freezing-point depression. For instance, for lysozyme approximately 0.32 g water/g protein remains in the liquid state down to temperatures below $-60\,^{\circ}$C. This fraction was considered to be equivalent to the amount of water required to satisfy all the exposed polar groups. The non-freezing phenomena has been observed with a variety of experimental techniques but the reasons for its existence are poorly understood. Evidently this water is unable to form a three-dimensional ice crystal structure but is able to retain similar rotational and translational properties to bulk water at temperatures well below normal freezing.

The most important hydration-induced property, and of central interest to hydration studies, is the onset of biological activity. A dry protein is completely inactive but it has been shown that activity is restored at a relatively low hydration, and certainly well below solution conditions. For example, lysozyme begins to show signs of activity at about 20 wt.% water. There may be a number of reasons for the onset of activity at this particular hydration level: the enzyme may require this amount of water to finally achieve the correct conformational state, the substrate might only find access to the active site possible, or the enzyme might only achieve the required kind of internal vibrational motion above this level of hydration. The rate of activity of lysozyme at 0.2 g protein/g water is very low but increases rapidly with hydration up to 0.5 g/g after which it follows a logarithmic rise to solution conditions.

DNA HYDRATION STUDIES

From the early X-ray diffraction studies it was clear that DNA has the ability to undergo well-defined structural transitions. Such transitions were found to be induced simply by changing the relative humidity of the DNA environment or by changing the counter-ion type or concentration. The base-pair sequence was also found to be crucial in determining the local conformation.

A relatively large number of dielectric studies have been carried out on DNA solutions and these have been primarily concerned with investigation of the origins of the two major high-frequency dielectric losses. This work has been discussed in Chapter 4. Studies of the water-binding properties of DNA have employed hydrated fibrous samples. In one such study, the dielectric properties of hydrated DNA at a fixed frequency of 10 GHz were used to probe the non-freezing fraction of DNA-bound water. The microwave cavity perturbation technique employed had the advantage of not requiring electrode contacts to the DNA sample. Measurements made at 10 GHz are able to detect the dielectric loss due to water molecules which are bound to DNA but which remain relatively rotationally unhindered and therefore exhibit a rotational relaxation time which is comparable with that of normal bulk water. By measuring the dielectric loss in this frequency region, it was possible to follow the changing rotational properties of the DNA-bound water with decreasing temperature and to observe the temperature at which this water froze (ice exhibits a dielectric loss some six orders of magnitude lower in frequency than water and shows a negligible loss in the gigahertz region). By monitoring the dielectric loss as a function of decreasing temperature, it was observed that the hydrated DNA samples exhibited a considerable dielectric loss well below the freezing point of normal bulk water as a result of the inability of the DNA-bound water to assume ice-like properties. For each DNA-hydration value, the dielectric loss was observed to fall to a level equivalent to that of 'dry' DNA at a well-defined temperature. The dielectric loss versus temperature characteristics indicated a gradual reduction rather than a sudden transition in the rotational mobility of the water molecules in the hydration shell of DNA as the temperature was lowered. From these freezing-point depression measurements it was concluded that approximately 280 water molecules per helix turn are perturbed by the DNA structure. This compares well with the figure of 260 water molecules/helix turn which were detected as unable to take up an ice-like structure in infrared studies.

LIPID AND MEMBRANE HYDRATION

The water-binding properties and hydration-induced changes in lipid bilayer structures have been the subject of a number of studies using dielectric techniques. In one such study, the frequency range from 4 to 12 GHz was employed to characterize the dielectric losses from the water molecules bound by the lipid head groups and from the very small dipoles ($\mu \sim 0.1$ Debye) associated with the end segments of the acyl chains. Of particular interest were the changes in bilayer lipid structure which occur at the lipid phase-transition temperature. These structural changes involve the transition from the rigid, *all-trans*, to the fluid, partially *gauche*, state of the fatty acyl chain. The variation of the permittivity of hydrated dimyristoylphosphatidylcholine (DMPC) bilayers as a function of temperature was found to exhibit a marked positive step discontinuity at the main transition temperature as shown in Fig. 5.5. This discontinuity increased in magnitude with increasing water content and the temperature range over which the response was

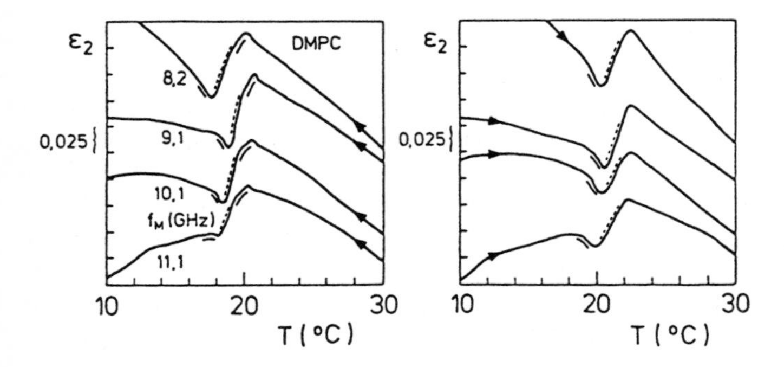

Fig. 5.5 Variation of permittivity with temperature in the region of the phase transition at four frequencies. (*Left*) cooling run (f_M = measurement frequency) and (*right*) heating run. Positive steps associated with chain melting are marked by broken lines and negative steps indicating water release and binding are underlined. The sample contains 32 water molecules/DMPC molecule.

From Enders, A. and Nimtz, G. (1984). Dielectric relaxation study of dynamic properties of hydrated phospholipid bilayers, *Ber. Bunsenges. Phys. Chem.* **88**, 512–517.

observed decreased markedly with increasing hydration. The hydration dependence of the transition characteristic probably reflects the ability of bound water molecules to elicit a higher degree of fluidity and cooperative rotational mobility in the higher temperature molten phase. Certainly the lateral area of the molten phase is known to increase with the amount of bound water indicating a decrease in the interactions between the polar head groups of the lipid molecules. In addition to the large positive step in permittivity at the main transition temperature of the bilayer, a small negative discontinuity was also apparent which preceded the positive response. As shown in Fig. 5.5, this feature was observed in both heating and cooling experiments and was thought to involve a change in the hydration of the polar head groups. An important conclusion of these observations is that the phase transition of the bilayer may be triggered by a change in the number of bound-water molecules and not, as commonly thought, by chain melting. The changes in the properties of the bound water are probably indicative of a conformational rearrangement of the head groups. It is likely that these conformational changes play an important role in initiating the phase transition process.

Some indication as to the state of the lipid-bound water is also provided by dielectric measurements. In a dielectric study in the gigahertz frequency range on hydrated dipalmitoylphosphatidylcholine (DPPC) bilayers, it was found that for hydration levels up to 15 wt.% (seven water molecules per lipid) the imaginary component of the permittivity had a value close to zero indicating that at these hydration levels, the lipid-bound water is considerably rotationally hindered with a relaxation frequency considerably lower than 1 GHz. For hydration levels >15 wt.% the contribution to the dielectric loss from the acyl chain segments was found to increase with water content probably reflecting a similar dielectric screening effect operating on the [head group–head group] interactions to that described for protein side chains. The increase in bilayer area with increasing water content elicits greater freedom to the acyl chains resulting in larger effective dipole moments.

In addition to the characteristic pre- and main transitions, a third structural phase transition has also been identified for DPPC at sub-zero temperatures using dielectric methods. This transition was attributed to large changes in head-group hydration which resulted in structural changes in the bilayer. It was envisaged that at the transition temperature, water molecules loosely associated with the phospholipid, i.e. those >15 wt.%, are squeezed out of the layer and then freeze along with the bulk water. The effect of this is to increase interactions between phospholipid molecules resulting in a 10% decrease in the lamellar repeat distance.

The hydration characteristics of more complex membrane systems such as *Artemia* cysts have also been studied using dielectric spectroscopy. These are composed of an inner mass of approximately 4000 cells surrounded by an acellular shell and are able to undergo hydration/dehydration cycles without loss of viability. Analysis of dielectric measurements over the frequency range 0.8–70 GHz suggest that approximately 1 g water/g *Artemia* cyst exhibits dielectric behaviour different from that of normal water.

Protein fluctuations

THE DYNAMIC NATURE OF PROTEIN STRUCTURES

In the past proteins were thought of as having static, rigid structures with enzymes and substrates interacting in a simple 'lock-and-key' or jigsaw configuration. Even allosteric proteins, where large conformational changes were known to be induced by ligand binding, were treated as proteins in which there were two possible static conformational states. In more recent years, there has been a gradual acceptance of a more dynamic picture of protein structure. Evidence of considerable structural motions has come from X-ray crystallography where poorly-resolved parts of the protein structure have indicated motional freedom, and from solvent accessibility studies where solvent exchange has been observed in parts of the internal protein structure which, from static packing considerations, should be solvent inaccessible. High-resolution X-ray structural studies on myoglobin have shown that no static pathways exist by which ligands such as oxygen or carbon monoxide could gain access to the haeme-binding site from the protein surface. Fluctuations of the protein structure must therefore be involved in the entrance and exit of these ligands from the protein interior.

It is interesting to note that substrate binding often has a marked effect on protein structural motions. Solvent-exchange studies carried out on such protein–ligand associations as trypsin/trypsin inhibitor and lysozyme/*N*-acetylglucosamine have indicated a significant decrease in the extent of structural fluctuations compared with the protein and ligand as free entities. In these cases the effect has not been confined to the immediate area of the protein–ligand association but has extended over a large proportion of the protein. It appears, therefore, that the bound substrate elicits an influence which is not limited to simply stabilizing the region of the active site.

Since standard X-ray diffraction studies are unable to provide direct information about protein motions, the protein structures described by X-ray crystallography can only be regarded as time-averaged structures. It is likely that there are a range of conformational states with similar energies determined by cooperative and correlated fluctuations about that average. This follows from the fact that protein folding is stabilized by weak interactions which are not much more energetic than thermal motions, so that the various interactions determining conformation may, in some proteins, be so finely balanced that a number of similar conformations may co-exist. Evidence for the existence of alternative protein conformations comes from X-ray crystallographic studies where a number of alternative protein structures have been observed for the same molecular unit in the same crystal structure and also for different crystallizing conditions. Since comparisons between crystal and solution properties have, on the whole, shown a high degree of similarity, these different structures observed in crystal studies may be examples of variations which exist in solution. A more concise description of the extent of conformational fluctuations has come from an X-ray diffraction study on ferrimyoglobin. Frauenfelder and co-workers have investigated changes in the conformation of this protein over the temperature range 250 to 300 K and have

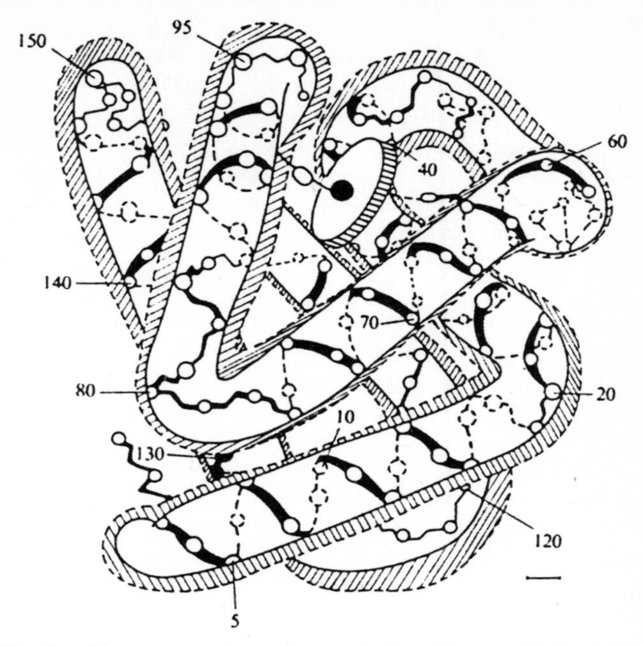

Fig. 5.6 The backbone structure of myoglobin. The solid lines indicate the path of the polypeptide chain: alpha-carbons are circles. Shaded areas are regions in which the polypeptide chain can move. Scale bar is 2 Å.
From Frauenfelder, H., Petsko, G.A. and Tsernoglou, D. (1979). Temperature-dependent X-ray diffraction as a probe of protein structural dynamics, *Nature* **280**, 558–563. Reprinted by permission from *Nature*, copyright 1979, Macmillan Magazines Ltd.

produced a picture, shown in Fig. 5.6, of the extent to which each part of the protein structure is able to fluctuate from the static average. Other X-ray studies on ribonuclease and lysozyme and denaturation studies on a number of enzymes have shown that for these enzymes the active site region is the area which supports the largest amplitude structural fluctuations.

The nature of the atomic bonding is crucial in determining the motional freedom of individual atoms. For instance the π-bonding in the aromatic rings of a number of side chains and in the peptide backbone will restrict the rotational freedom of the groups involved. In addition, there will often be steric considerations. For instance, steric restrictions to side-chain motion will result from collisions with other side chains and with the peptide backbone. In fact from packing considerations, in many cases it is likely that concerted side-chain motion is the rule rather than the exception.

The scales of protein internal motions in terms of amplitude, energy and time vary over many orders of magnitude indicating a range of possible individual, collective and cooperative types of motions. For instance, small, closely-packed groups of atoms may experience very short-time, small-amplitude motions in the order of 10^{-13} s and 0.2 Å, respectively. In less closely-packed regions of the protein, much larger displacements may occur, possibly involving a partial unfolding of a region of the protein. The types of motion involved may include rotational orientation of groups about individual bonds, the concerted movement of a number of side chains or section of the main peptide chain relative to the rest of the protein structure, or the fluctuating motion of counter-ions or solvent molecules associated with the protein.

THE INFLUENCE OF BOUND WATER ON PROTEIN DYNAMICS

One of the thermodynamic considerations of protein dynamics is the flexibility of the protein structure and implicit in a discussion of protein structural flexibility is the influence of *protein-bound water*. In recent dielectric studies on hydrated lysozyme and chymotrypsin the influence of the electrostatic interaction between the enzyme and bound solvent was discussed. In the study on hydrated chymotrypsin, it was noted that the magnitude of the dielectric loss for the lower frequency dielectric dispersion, attributed to protein structural motions, was relatively insensitive to water content up to hydration levels of ~160 water molecules per chymotrypsin as depicted in Fig. 5.4. It is only at hydration levels above this that a significant dielectric loss and therefore mobility in the protein structure occurs. This observation is consistent with the scheme in which protein-bound and structured water molecules are instrumental in dielectrically screening electrostatic interactions between polar species in the protein structure, thereby eliciting greater freedom of motion. Water molecules occupying the primary sorption sites (up to ~90 water molecules per protein) and those in the transition region (from 90 to 160 water molecules per chymotrypsin) have restricted rotational mobility and hence possess a much reduced polarizability and can be expected to provide only limited dielectric screening. The water bound in excess of 160 molecules per protein, however, appears to be able to exercise a considerably greater rotational polarizability and is therefore capable of significantly reducing the energies of interaction between the charged and polar groups in the protein structure. The fact that the dielectric loss associated with the protein flexibility continues to increase with increasing water content at relatively high hydration values (>400 water molecules per chymotrypsin) would seem to indicate that the measurements reflect a general loosening of the protein structure and not only an increase in the flexibility of the ionizable side chains.

It appears then that loosely-bound water molecules, i.e. those assumed to be bound by single hydrogen bonds, possess a relatively high dielectric permittivity and are therefore able to act as a plasticizer to the protein structure. It is envisaged that this type of hydration-dependent screening may allow the protein specific vibrational modes discussed above, and that these could be important in determining the protein's biological activity. This scheme is supported by the observation that a significant increase in the motional freedom of the protein structure occurs at hydration levels approaching that at which the onset of the enzyme activity of lysozyme and chymotrypsin has been observed. This is also consistent with electron spin resonance spin probe measurements on lysozyme which have indicated a direct correlation between the motional freedom of the lysozyme structure and the extent of enzymatic activity.

THE RELATION BETWEEN PROTEIN DYNAMICS AND ACTIVITY

As discussed above, one of the most important aspects of protein dynamics is the possibility that certain vibrational modes of protein structures facilitate biological activity. The possibility that such fluctuations play a part in, for instance, enzyme action may go some way to explaining a long-standing fundamental question regarding the relative smallness of the enzyme's active site compared with its total volume. The previously held view that the total enzyme structure is required to form and maintain the three-dimensional conformation of the active site is now in doubt. An alternative explanation might be that large enzyme structures are required to transmit, in a very specific way, the translational energy of solute molecules to the enzyme's active site. Of the many possible fluctuations of the enzyme structure, some may be of a correlated, directional nature which serve to provide the optimum energetic conditions at the active site required to produce the transition state and enzyme activity.

Careri and co-workers have suggested a model of enzyme action which involves dynamic coupling of the thermal energy of the ambient medium to the catalytic events at the active site. They have assumed that this type of coupling should be apparent in the fluctuational behaviour of the protein and have estimated that coupling of the solvent's thermal energy to the catalytic process requires events at the protein–solvent interface on a 10^{-8} s time scale. In fact a number of different dynamic processes associated with the enzyme surface such as proton-transfer reactions, local conformational motion, counter-ion fluctuations and bound water relaxations, take place around this time scale. This is thought to indicate the likelihood that the solvent–protein interface and the active site are statistically coupled and that free energy can be exchanged between the environment and the enzyme through coupling events at the protein surface. This model has been extended to consider enzymes as being composed of domains which, on the time scale of interest ($\sim 10^{-8}$ s), are themselves internally relatively rigid structures. It is envisaged that each of the domains have some degree of freedom and are able to interact with the substrate with one or two critical residues per domain at the active site. In this model domain rigidity is important in reducing the number of degrees of freedom of the enzyme, so reducing the search time for the optimum active-site conformation. Thus, a large number of small solvent interactions at the protein surface are thought to control the displacement of the catalytic residues at the active site. In this scheme internal and surface-bound water are likely to be important in determining the position and boundaries of the rigid domains.

Dielectric studies of ionic and protonic conduction in proteins

Protonmotive forces are the central energy-producing phenomena in biology. It is now generally accepted that protons are responsible for coupling electron flow through the electron transport chain to ATP synthesis although there is still some

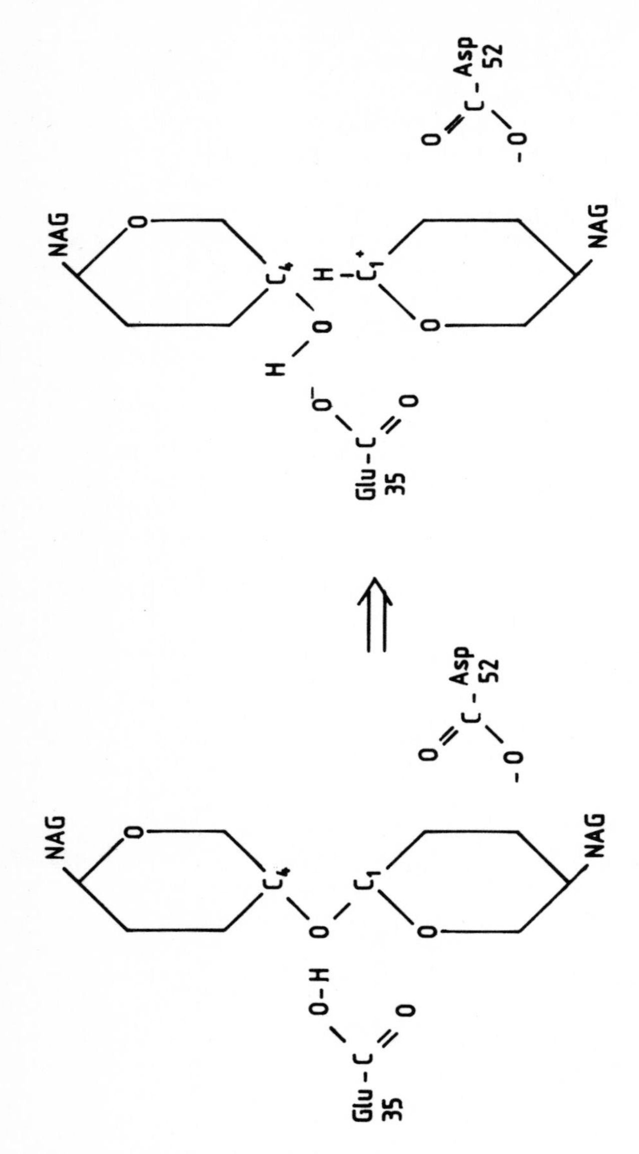

Fig. 5.7 Proton transfer in lysozyme during general acid catalysis. NAG = n-acetylglucosamine.

question as to whether the protons are delocalized in the aqueous medium or localized at the membrane surface. Proton gradients also provide energy for such diverse processes as active transport, NADPH synthesis, flagellar motion and heat production. In addition, enzymatic mechanisms often involve facilitated proton transfer between enzyme and substrate. For instance, in the enzymatic mechanism proposed for lysozyme, a proton is transferred from an un-ionized glutamic acid residue to a glycosidic oxygen atom 3 Å away as shown in Fig. 5.7. The glycosidic bond is cleaved as a result and a transient carbonium ion intermediate formed. This reaction is known as general acid catalysis and is common to the activity of a number of enzymes.

A mechanism involving more long range proton transport has been proposed for a number of proteolytic enzymes such as chymotrypsin. The essential catalytic activity of chymotrypsin depends on the ability of three specific hydrogen-bonded side groups in the vicinity of the active site to form a charge relay network and act as a proton shuttle during the hydrolysis of the substrate as illustrated in Fig. 5.8.

In addition to these well-documented cases, a more general role for protons in enzyme action has been proposed. In the active centres of many enzymes there are a high proportion of ionizable groups capable of proton transfer and it has been postulated that the three-dimensional structure of the enzyme molecule may be responsible for connecting the aqueous environment to the active site electrically through proton networks. Such proton pathways might take the form of regular α-helices, β-sheets or extended hydrogen-bonded internal water structures. Local electric field effects which occur as a result of the removal of protons from one end of a proton pathway may be transmitted through the network to produce an electric field effect a considerable distance away. This type of extended proton transfer through hydrogen-bonded chains may play an important role in enzymatic activity.

Low-frequency dielectric and conduction studies of hydrated protein powders have been carried out in a number of laboratories over the last thirty years and a fairly consistent picture of their electrical properties has emerged. Two main dispersive losses are observed in the low-frequency region and these have been designated the α and Ω losses as illustrated in Fig. 5.9. The Ω-dispersion takes the form of a large featureless loss which increases in magnitude and hence increasingly dominates the lower frequency spectrum with increasing hydration. The α-dispersion, on the other hand, is more well defined and exhibits a hydration-dependent relaxation time which typically decreases by nine orders of magnitude for an increase in water content from a dry to a fully-hydrated state (approximately 0.4 g water/g protein). The magnitude of the α-dispersion is approximately independent of hydration.

A similar dramatic change in the d.c. conductivity is also observed with increasing hydration, as illustrated in Fig. 5.10(a) for the enzyme lysozyme. Analysis of the electrical properties of a number of hydrated proteins has shown the d.c. conductivity to be related to the relaxation time of the α-dispersion by the equation

$$\tau = \varepsilon_0 \varepsilon_\infty / \sigma \tag{5.2}$$

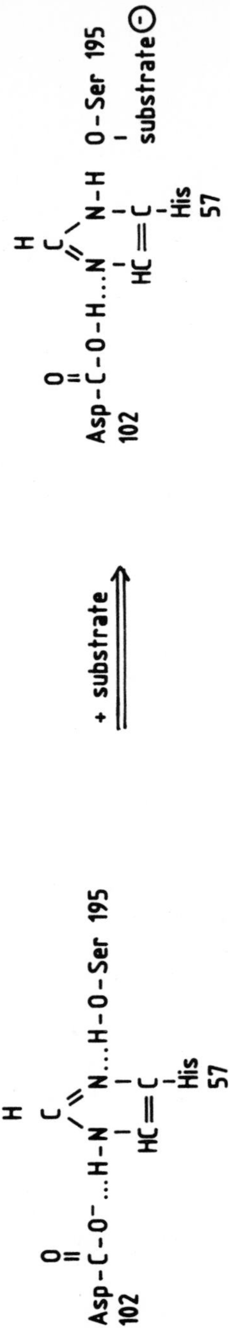

Asp-C(=O)-O$^-$...H-N (His 57 imidazole: N, CH, N, HC=C) N...H-O-Ser 195

Asp 102 · His 57 · Ser 195

Asp-C(=O)-O-H...N (His 57 imidazole: N, CH, N, HC=C) N-H   O-Ser 195 | substrate$^{\ominus}$

Asp 102 · His 57 · Ser 195

Fig. 5.8 The proton shuttle in chymotrypsin.

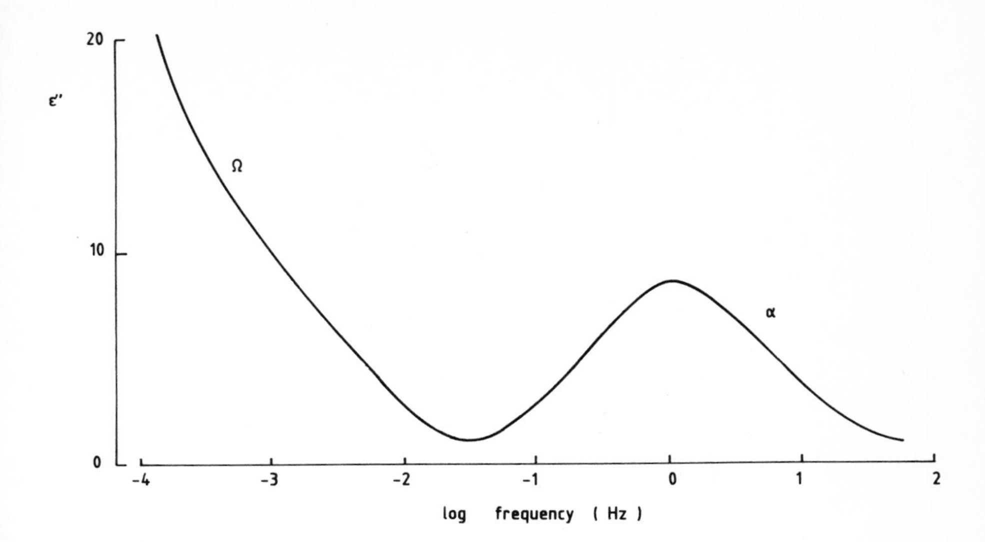

Fig. 5.9 Low-frequency dielectric loss characteristic for a typical hydrated protein.

where ε_0 is the permittivity of free space, and ε_∞ is the relative high frequency permittivity of the α-dispersion.

In spite of these numerous studies, there is still considerable discussion regarding the type of charge carriers responsible for the observed conductivity, the conduction mechanism involved and the origins of the dielectric dispersion. Some important experimental evidence has recently provided some insight into these matters. Using normal copper or gold electrodes, discoloration at the sample–electrode interface is observed after the passage of d.c. current, the effect being particularly noticeable at high hydrations. When blocking electrodes of PTFE (polytetrafluoroethylene) are interposed between the copper electrodes and the protein sample to reduce charge carrier injection into the protein, the **Ω**-dispersion disappears or is at least greatly reduced in magnitude. From this evidence it seems likely that the **Ω**-dispersion can be attributed to the accumulation of ionic species at the sample–electrode interface. These ions may be associated with the protein as counter-ions or arise from the buffer solution employed prior to lyophilization.

In the case of the α-dispersion, the experimental evidence is not so clear. Some indication as to the origins of this dispersion can be deduced from considering the magnitude of the dielectric decrement. From the standard dielectric theories discussed in Chapter 3 it has been shown that the dielectric decrement, $\Delta\varepsilon$, is approximately proportional to $\mathcal{N}<\mu>^2$, the product of the number density of the dipolar species and the square of the effective dipole moment. One of the possible mechanisms responsible for the α-dispersion is rotation of the polar entities present

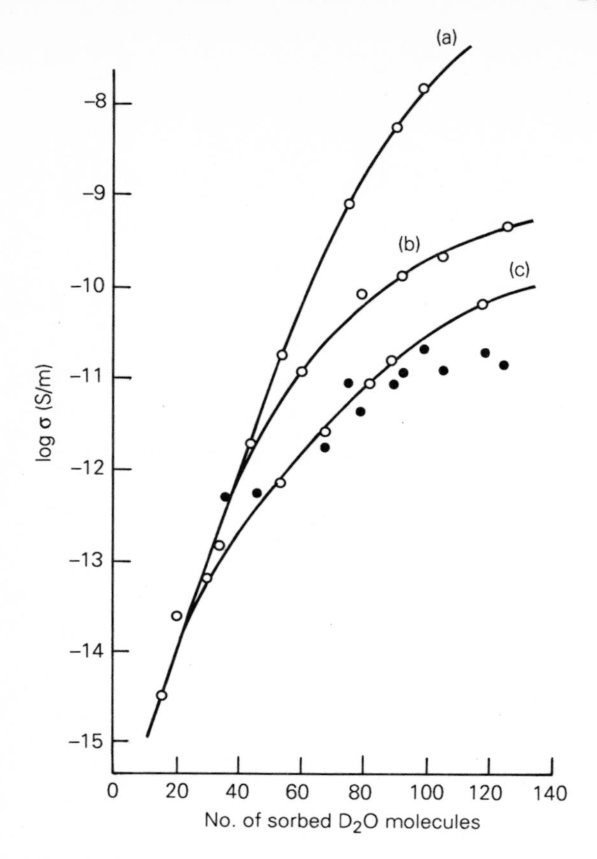

Fig. 5.10 Variation of the conductivity of deuterated lysozyme as a function of D_2O content for: (a) untreated lysozyme, (b) 240 h field cleaning and (c) 500 h field cleaning. (●) Deuterium ion conductivity determined using solid-state electrolysis measurements.
From Morgan, H. and Pethig, R. (1986). *J. Chem. Soc. Faraday Trans. I* **82**, 143–156.

in the protein structure such as the peptide units and polar side chains. It turns out, however, that the magnitude of the α-dispersion is such that for the α-dispersion to be interpreted in terms of this type of process, each peptide unit would need to exert its full dipole moment. This would require each of the peptide units to be free to rotate in the protein structure. Such rotational freedom is unlikely for solid hydrated proteins, and a dipolar orientational process would, therefore, seem to be discounted. The processes which seem to be the most likely candidates in this case are Maxwell–Wagner interfacial polarization or charge-carrier hopping. The former might be associated with polarization due to charge carriers accumulating at grain boundaries in the bulk of the sample. In this case the effect of hydrating

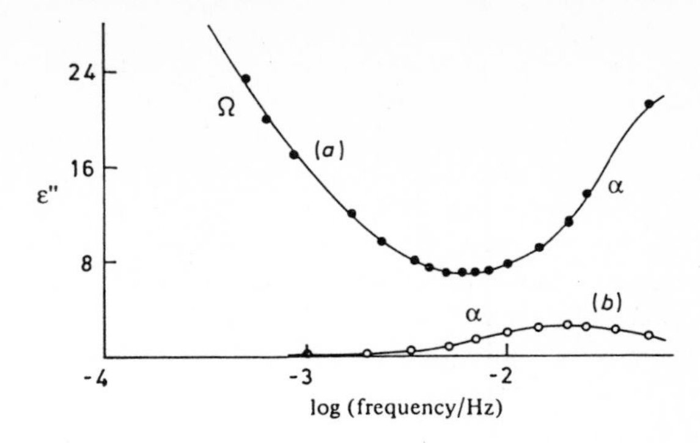

Fig. 5.11 Frequency dependence of the dielectric loss ε'' for lysozyme at 11% water content. (a) Untreated lysozyme, (b) field-cleaned lysozyme.
From Morgan, H. and Pethig, R. (1986). Protonic and ionic conduction in lysozyme, *J. Chem. Soc. Faraday Trans. I* **82**, 143–156.

the sample would be to alter the conductivities and permittivities of the different phases producing a hydration-dependent change in the relaxation time and total conductivity. In the latter case, ions and/or protons might be involved in hopping motions between polar and ionized groups in the protein structure. The hydration dependence of the conductivity and relaxation time in this case could be explained in terms of the ability of protein-bound water to electrostatically screen the charge carriers from the opposite charges on the protein, effectively reducing the local potential energy barrier associated with the hopping process.

Since there is a direct relationship between the relaxation time of the α-dispersion and the steady-state conductivity and since the activation energies for both processes are similar, it seems likely that the same conduction mechanism may be responsible for both processes. So, for instance, if the α-dispersion were attributed to local charge-carrier hopping, the d.c. conductivity might be expected to be a result of extended long-range ion hopping through the sample along pathways which connect local sites.

Another important aspect of the conduction properties of hydrated proteins is the type of charge carrier involved. Again the relationship between τ and σ (equation 5.2) would seem to indicate that the same charge carrier could be responsible for both a.c. and d.c. processes. Some recent experimental evidence involving the study of 'field-cleaned' protein samples is relevant to this question. Field cleaning involves the application of very high electric fields to hydrated samples which results in the electrophoretic removal of excess ions from the bulk of the sample. After the electric field has been applied for up to 3 days, the accumulated ions at the electrode–sample interface are removed and the protein sample is then considerably more free of ions than if it had been subjected to extensive conventional dialysis against ultra-pure water. It should be noted that, despite the apparent harshness of this procedure, the enzyme remains in a fully active state. Field cleaning is found to dramatically alter both conduction and dielectric properties of proteins as well as their hydration characteristics. As shown in Figs. 5.10 and 5.11, the relaxation time of the α-dispersion is considerably increased and its magnitude reduced and the d.c. conductivity is also reduced, these effects being most noticeable at high hydrations. These experiments seem to indicate that, for protein samples containing excess ionic species, the charge carriers are ions. The α-dispersion may be attributed to local hopping of these ions and conductivity determined by long-range migration of these through the sample. In the case of proteins which have been dialyzed against ultra-pure water, it seems likely that ionic conduction is still dominant, the major charge carriers here being counter-ions. Only after employing the more rigorous field-cleaning technique does the ionic conductivity fall to the level where a different charge carrier is observed.

The nature of the charge carrier in field-cleaned samples is indicated by recent conduction measurements on deuterated proteins and by solid-state electrolysis measurements on field-cleaned lysozyme. Figure 5.12 shows the marked fall in d.c.

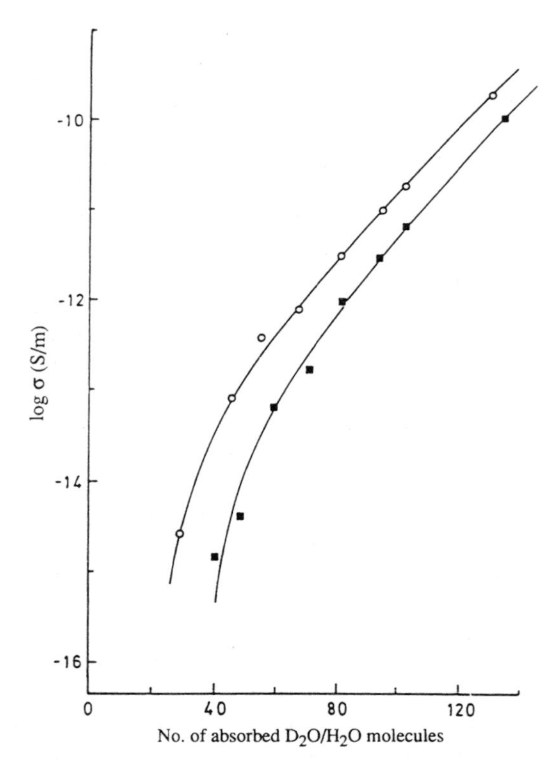

Fig. 5.12 Conductivity of normal (○) and deuterated (■) field-cleaned lysozyme as a function of H_2O and D_2O content, respectively.
From Morgan, H. and Pethig, R. (1986). Protonic and ionic conduction in lysozyme, *J. Chem. Soc. Faraday Trans. I* **82**, 143–156.

conductivity when D_2O is absorbed by the sample compared with normal hydration with H_2O. In addition, measurements have been made of the number of deuterium molecules released by electrolysis at the cathode during the passage of a known amount of charge through a deuterated protein sample. These experiments have led to the conclusion that the dominant charge carriers in hydrated field-cleaned proteins are protons. It is envisaged that the protons originate from the ionizable groups associated with the protein structure and that they are able to hop from site to site along a network of water–protein and water–water hydrogen bonds. In such a scheme the effective proton mobility depends upon the number of water-based percolation pathways on the surface of the protein molecule. In the case of the enzyme lysozyme, there is some evidence to suggest that the binding of the substrate (*N*-acetylglucosamine) increases the magnitude of the α-dispersion. This may be an important observation perhaps indicating proton interaction with the substrate at the active site. However, further work is necessary to eliminate the effect of counter-ions and to definitely identify the α-dispersion with a particular charge carrier. In any case, the important point is that proteins are able to support long-range proton transport although many of the studies conducted to date may have over-estimated its magnitude due to contributions from other ionic species.

In addition to ionic and protonic conduction in protein structures, electronic transport also plays an important role in the function of a number of proteins. In a study of the conduction properties of the protein cytochrome c_3, an electron carrier in the sulphate-reducing bacteria *Desulfovibrio vulgaris*, measurements indicated that the protein's conductivity is greatly dependent on its oxidation state. The fully-reduced ferrocytochrome c_3 exhibits an extremely high d.c. conductivity ($\sim$0.13 mho/cm) which approaches that observed for semi-metals. The conduction mechanism envisaged here is one of electron hopping governed by intermolecular haeme–haeme interactions.

Transmembrane proton and ion transport

In this section we turn to a discussion of proton and ion transport across membranes with particular reference to the role of protons in the energy-transduction process. It is generally accepted that in ATP synthesis protons migrate through the F_o component of the ATPase complex which spans the inner membrane of mitochondria. As previously mentioned, some question still remains as to whether the protons are delocalized in the bulk aqueous phase or more localized at the membrane surface. In fact, other fundamental questions also remain unanswered concerning the protonic conduction mechanism and the molecular entities which constitute the proton pathway.

A number of models have been proposed describing mechanisms for proton transport across the mitochondrial membrane. A proton could be attached to a carrier molecule on one side of the membrane, transported across and released on the other side. Another possible mechanism involves a large conformational change in the structure of a membrane protein which effectively transports a proton associated with the protein across the membrane. Alternatively, the

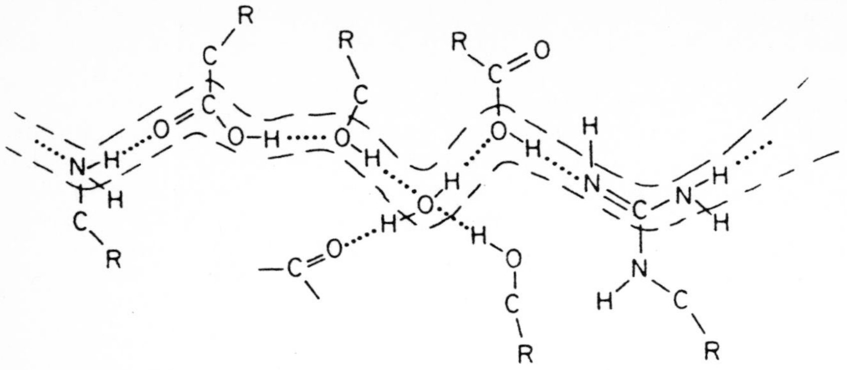

Fig. 5.13 Hydrogen-bonded chain formed from amino acid side chains of membrane proteins and some bound water molecules.
From Nagle, J.F. and Tristram-Nagle, S. (1983). Hydrogen bonded chain mechanisms for proton conduction and proton pumping, *J. Membrane Biol.* **74**, 1–14.

protons might be transported through the membrane via proton channels existing within the membrane protein structures. The proton-channel model has been developed in publications by Nagle and co-workers. They envisaged the existence of proton pathways constructed from extended hydrogen-bonded chain structures formed between polar amino side chains and bound water molecules as depicted in Fig. 5.13. The side chains are located in the required hydrogen-bonding positions in the channel by the membrane protein. The conduction mechanism envisaged in this proton-transport model is identical to that proposed for protonic conduction in ice, i.e. the so-called *hop-turn process*. A positive ionic defect (H_3O^+) is propagated through the crystal structure followed by the rotation of the hydroxyl groups back along the hydrogen-bond chain which propagates a second negative L defect in the opposite direction as shown in Fig. 5.14. A linear arrangement of approximately 20 side chains capable of hydrogen bonding are necessary to span the membrane. Such extended hydrogen bonding has been shown to be theoretically possible in β-sheeted structures and in parallel α-helical arrangements. Unfortunately, although the amino acid sequences of a number of membrane proteins have been elucidated, to date no high-resolution X-ray structural information is available. There is therefore no direct experimental evidence for the existence of such proton pathways. Despite the lack of structural detail, however, the likely arrangement of side chains within the channel would seem to favour the formation of hydrogen bonds since it is energetically favourable for the polar groups of membrane proteins to be sequestered in the protein interior while the non-polar, hydrophobic side chains face the membrane lipids. For a viable proton pathway to exist however, the polar side chains must be capable of supporting a continuous hydrogen-bond chain across the membrane without a break.

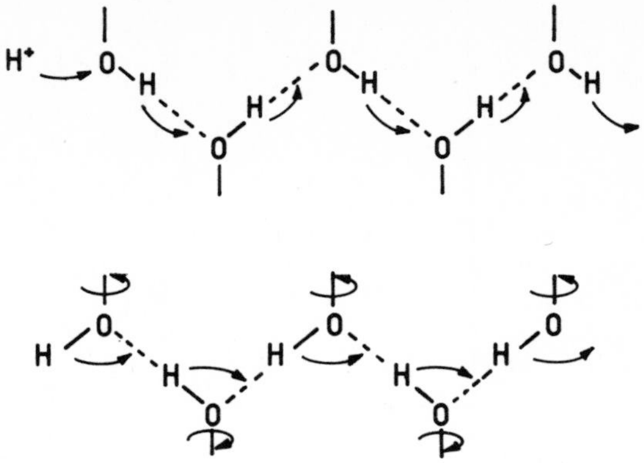

Fig. 5.14 Protonic conduction mechanism envisaged for ice structures involving migration of positive ionic and negative rotational defects.

An important example of a transmembrane protein through which protons are transported is the c-subunit of the F_0 component of the H^+–ATPase complex (also known as the DCCD-binding protein) which spans the inner membrane of mitochondria and which is active in oxidative phosphorylation. In oxidative phosphorylation, protons are pumped across the inner membrane (from the matrix side to the cytoplasmic side) and a proton gradient is built up in response to electrons passing sequentially down the electron-transport chain to electron carriers with increasing electron affinity. In the event of a need for ATP, protons flow back to the matrix side through a channel which is part of the F_0 component of the ATPase complex. The flow of protons back through the channel results in ATP synthesis. The chloroplast ATPase system in plants is similar to the mitochondrial system in that a proton gradient is generated, in this case by light energy.

We will now consider the likelihood of such a continuous hydrogen-bond chain system in the F_0 component. The predominantly hydrophobic nature of the transmembrane F_0 component can be clearly seen from the amino acid sequences of the protein from a number of widely different species (Table 5.1). It is also evident from these sequences that there is a high degree of homology both in the general character of the polypeptide chain organization and in the positions of some of the residues. Conservation of amino acids in a protein is usually indicative of the importance of the role of those particular amino acids in the function of the protein. It is likely therefore that many of the amino acids which actually define the channel are very specific, their physico-chemical properties being essential in determining the activity of the channel as a whole.

Possible arrangements of this protein within the membrane are described by Ovchinnikov and associates using amino acid sequences determined by Sebald and Wachter (Table 5.1). Only as few as 25% of the side groups are hydrophilic and there is clearly-defined clustering of the hydrophobic and hydrophilic residues. This type of clustering of polar and non-polar groups is energetically

Table 5.1 Amino acid sequences of the DCCD-binding proteins from *Neurospora crassa* (*N. cr.*), *Saccharomyces cerevisiae* (*S. cer.*) and *E. coli* (*E. c.*).

N. cr.	Tyr-	Ser-	Ser-	Glu-	Ile-	Ala-	Gln-	Ala-	Met-	Val-	Glu-	Val-	Ser-	Lys-	Asn-	Leu-	Gly-
S. cer.							f-Met-	Gln-	Leu-	*Val-*	Leu-	Ala-	Ala-	*Lys-*	Tyr-	Ile-	*Gly-*
E. c.					f-Met-	Glu-	Asn-	Leu-	Asn-	Met-	Asp-	Leu-	Leu-	Tyr-	Met-	Ala-	Ala-
N. cr.	Met-	Gly-	Ser-	Ala-	Ala-	Ile-	Gly-	Leu-	Thr-	Gly-	Ala-	Gly-	Ile-	Gly-	Ile-	Gly-	Leu-
S. cer.	*Ala-*	*Gly-*	Ile-	Ser-	Thr-	*Ile-*	*Gly-*	*Leu-*	Leu-	*Gly-*	*Ala-*	*Gly-*	*Ile-*	*Gly-*	*Ile-*	Ala-	Ile-
E. c.	Ala-	Val-	Met-	Met-	Gly-	Leu-	Ala-	Ala-	Ile-	*Gly-*	*Ala-*	Ala-	*Ile-*	*Gly-*	*Ile-*	Gly-	Ile-
N. cr.	Val-	Phe-	Ala-	Ala-	Leu-	Leu-	Asn-	Gly-	Val-	Ala-	Arg-	Asn-	Pro-	Ala-	Leu-	Arg-	Gly-
S. cer.	*Val-*	*Phe-*	*Ala-*	*Ala-*	*Leu-*	Ile-	*Asn-*	*Gly-*	*Val-*	Ser-	*Arg-*	*Asn-*	*Pro-*	Ser-	Ile-	Lys-	Asp-
E. c.	Leu-	Gly-	Gly-	Lys-	Phe-	Leu-	Gln-	*Gly-*	Ala-	Ala-	*Arg-*	Gln-	*Pro-*	Asp-	Leu-	Ile-	Pro-
N. cr.	Gln-	Leu-	Phe-	Ser-	Tyr-	Ala-	Ile-	Leu-	Gly-	Phe-	Ala-	Phe-	Val-	Glu-	Ala-	Ile-	Gly-
S. cer.	Thr-	Val-	*Phe-*	Pro-	Met-	*Ala-*	*Ile-*	*Leu-*	*Gly-*	*Phe-*	*Ala-*	Leu-	Ser-	*Glu-*	*Ala-*	Thr-	*Gly-*
E. c.	Leu-	Leu-	Arg-	Thr-	Gln-	Phe-	Phe-	Ile-	Val-	Met-	Gly-	Leu-	Val-	Asp-	*Ala-*	Ile-	Pro-
N. cr.	Leu-	Phe-	Asp-	Leu-	Met-	Val-	Ala-	Leu-	Met-	Ala-	Lys-	Phe-	Thr-				
S. cer.	*Leu-*	*Phe-*	Cys-	*Leu-*	*Met-*	*Val-*	Ser-	Phe-	Leu-	Leu-	Leu-	*Phe-*	Gly-	Val-			
E. c.	Met-	Ile-	Ala-	Val-	Gly-	Leu-	Gly-	Leu-	Tyr-	Val-	Met-	*Phe-*	Ala-	Val-	Ala-		

From Sebald, W. and Wachter, E. (1978). Amino acid sequence of the putative protonophore of the energy-transducing ATPase complex, in *Energy Conservation in Biological Membranes*, Eds. Schäfer, G. and Klingenberg, M. Springer-Verlag.

preferred and infers the kind of structural arrangement shown in Fig. 5.15, with the hydrophobic sections crossing the membrane, leaving the polar segments exposed to the bulk aqueous phase or making specific contacts with other ATPase subunits. The proton channel is thought to be defined by between six and twelve such protein arrangements. From the relatively small number of polar side chains per protein within the channel capable of hydrogen bonding, at first sight it would seem unlikely that a continuous hydrogen-bonded chain through the F_0 channel could exist. However Nagle *et al.* point out that if this number is multiplied by six (for each of the protein units which define the channel) then, in theory, it is possible for the polar side groups in the channel to provide a continuous hydrogen-bond chain across the membrane. A variation of this scheme is the flexible channel model in which several shorter proton chains are envisaged with gaps which are bridged transiently as a result of small conformational motions in the membrane protein structure.

An alternative approach to the protein-based hydrogen-bond chain may be found by considering the role that water molecules might play in transporting protons through transmembrane channels. It has been suggested that water in the channel might itself support sufficient proton transport to account for observed rates of ATP synthesis. The mobility of protons in water is approximately the same as in ice and therefore would be expected to be comparable with a hydrogen-bonded chain. However, a simple channel supporting normal bulk water would also conduct ions and might result in an excessively high electric field being applied to the active site of the ATPase complex. Ionic conduction of this kind would not occur in a narrow pore defined by the membrane protein, since, in the absence of bulk water, a large energy barrier exists for ions attempting to enter the channel in a fully hydrated state. If we consider now the ATPase F_0 channel and the nature of the water which might be located therein we find the predominant interaction between the membrane protein and the associated water to be primarily a hydrophobic one.

In the previous discussions of protein–water interactions we were mainly concerned with the binding of water to polar side chains and peptide units. However, as mentioned, in solution, and at high hydrations, exposed apolar hydrophobic groups will interact with the surrounding water molecules. It has been suggested that the water molecules surrounding hydrophobic groups and hydrophobic patches on protein surfaces may be partially stabilized in cage-like structures such as those known to exist around non-polar atoms in the clathrate hydrates. These hydrate cage-like water structures, an example of which is shown in Fig. 5.16, are known to form when non-polar solutes are dissolved in water. None is able to readily hydrogen-bond to water and all are sparingly soluble. The process of dissolving such apolar substances in water is accompanied by a significant decrease in entropy and this has been interpreted as a water 'structure-making' phenomenon. Since the non-polar solutes occupy space, the random hydrogen-bond water network must re-organize around it in such a way that sufficient room is available to accommodate the guest clathrate molecule while at the same time maximizing the number of water–water hydrogen bonds and avoiding too much damage to the hydrogen-bond network. The cage is usually far

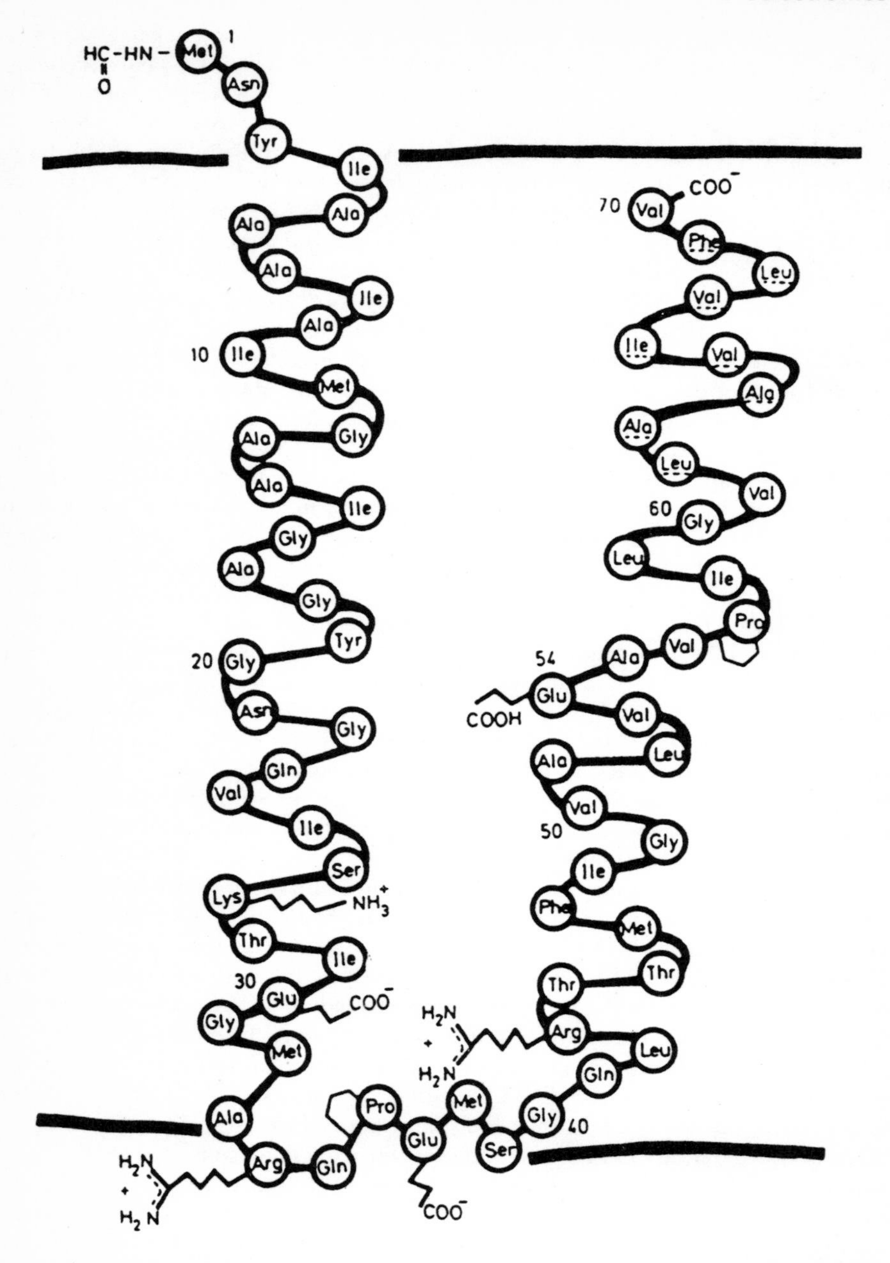

Fig. 5.15 Possible arrangement of the DCCD-binding subunit of H^+-ATPase. From Ovchinnikov, Yu. A., Abdulaev, N.G. and Modyanov, N.N. (1982). Structural basis of proton translocating protein function, *Ann. Rev. Biophys. Bioeng.* **11**, 445–463. Reprinted with permission from the *Annual Revue of Biophysics and Bioengineering*, copyright 1982 by Annual Reviews Inc.

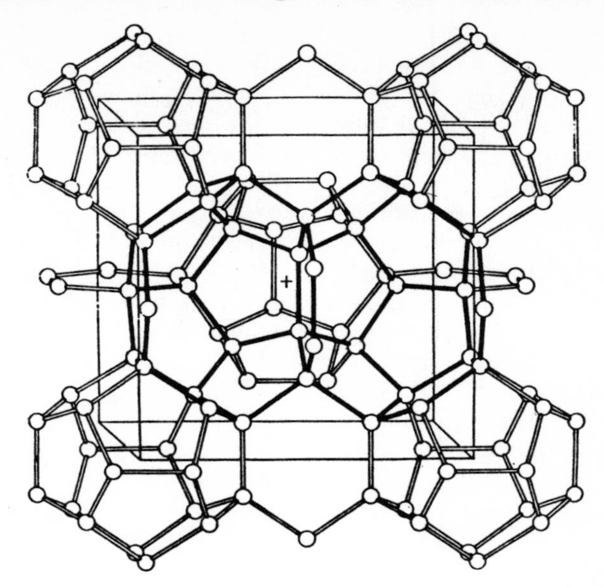

Fig. 5.16 The polyhedral hydrogen-bonded oxygen framework of a type I clathrate hydrate structure.
From McMullan, R.K. and Jeffrey, G.A. (1965). Polyhedral clathrate hydrates, *J. Chem. Phys.* **42**(8), 2725–2732.

from perfect and although there is some strengthening of hydrogen bonds compared to those present in bulk water, substantial strain and disorder remain. The resulting hydrogen-bonded structures are by no means solid or long-lived but exhibit significantly increased correlation times and lessened irregularity.

Experimental evidence for such water structures include dielectric studies employing very sensitive dielectric difference measurements in the range 8–25 GHz which have investigated the influence of hydrophobic solutes on the dynamic behaviour of water. In these investigations it was found that the dielectric relaxation rate of water was significantly reduced by the dissolution of hydrophobic solutes indicating increased hydrogen bonding between water molecules in the immediate vicinity of the hydrophobic solute. In addition, computer simulation studies of the structure of liquid water at extended hydrophobic surfaces has shown that this interaction produces density oscillations that extend at least 10 Å into the liquid and molecular orientational preferences that extend at least 7 Å indicating a more long-range ordering of the water molecules. Clathrate hydrate structures at temperatures below 0 °C have also been the subject of dielectric studies. These have indicated that hydrate formation is accompanied by a shift in the relaxation frequency of the H_2O lattice to higher values than that associated with ice. This may be consistent with a less rigid hydrogen-bonded structure being present for the hydrates compared with crystalline ice. A dielectric study of the largely hydrophobic transmembrane oligopeptide gramicidin incorporated into an ice matrix has revealed the existence of two relaxation peaks, one associated with the ice structure and a second at higher frequencies possibly due to the same kind of water hydrate structure observed for clathrate hydrates.

Returning now to the F_0 channel of the ATPase complex, the existence of water structures surrounding the large number of hydrophobic groups within this channel would seem to be relevant to proton transport. If, as seems likely, there is an increase in hydrogen bonding of the water due to the hydrophobic nature of the protein side groups this would result in a significant increase in the ability of those water molecules to conduct protons through the transmembrane channel. The protein channel would, in this scheme, be responsible for 'structuring' the water, with this structured water supporting proton transport through the channel. Compared with the rigid crystal structure of ice, this type of structured water would be more relaxed with water molecules free to arrange themselves rather than being constrained by symmetry requirements.

We can now consider the question of the likely conduction mechanism through the channel. There has been the tendency to consider the proton-transport mechanism proposed for ice as being the most relevant to proton transport in biological systems. However, this may not be the most appropriate model. The hydrogen-bonding systems present in biological molecules tend to be far more disorganized than those existing in the rigid crystalline ice configuration. A more relevant hydrogen-bond model might therefore be one existing in a more disordered system. Such disordered hydrogen bonding exists in a class of carbohydrates known as the cyclodextrin hydrates. The existence of extended networks of water-based circular hydrogen bonds has been shown in α-cyclodextrin hydrate and neutron diffraction studies on β-cyclodextrin hydrate have shown extended hydrogen-bond networks in which a proportion of the hydrogen bonds are of the so-called 'flip-flop' type. In these hydrogen bonds, the two possible H locations (O–H= = =H–O) are mutually exclusive. The result is that only one of the two H atoms can be in hydrogen-bonding contact at a given time and the other H is flipped out to form a hydrogen bond with an adjacent oxygen atom. This phenomena has resulted in cooperative and concerted motions in the hydrogen-bond chains and the apparent domino effect shown in Fig. 5.17. Conduction studies employing hydrogen-saturated palladium black contacts as proton-injecting electrodes on α-cyclodextrin hydrate have shown protonic conductivities some two orders of magnitude greater than that detected for ice and this may prove to be a considerable underestimate when the intercrystalline grain boundary resistances of the cyclodextrin samples used are taken into account. A comparable hydrogen-bonding scheme has been found in ice clathrates. The major difference here is that the oxygen positions are constrained by symmetry requirements whereas the structure is more relaxed and therefore disordered and the oxygen positions are freely arranged in the cyclodextrin hydrates as they are in biomacromolecules.

A similar freedom to arrange may be apparent in the water structure existing in the vicinity of hydrophobic side groups and it is possible that an extended hydrogen-bond system exists in the water 'structure' inside the F_0 channel defined

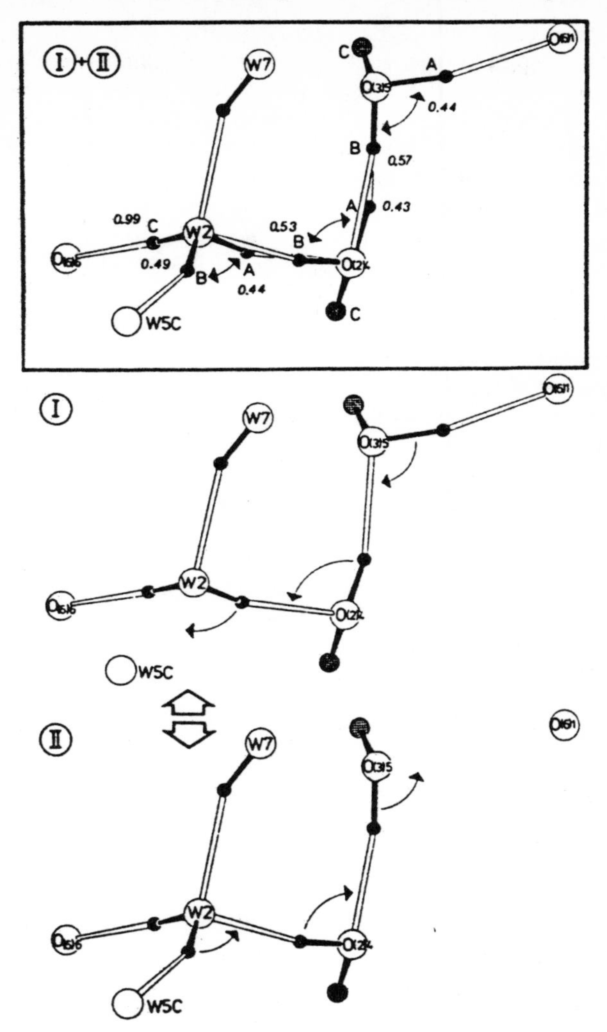

Fig. 5.17 Deconvolution of the flip-flop hydrogen-bonding system into two states (I and II) having normal hydrogen bonds of the type O–H······O. Solid arrows indicate the manner in which the H atoms may jump in concerted, cooperative motion from one state to another. Water molecule W2 displays three H-atom positions, one (C) fully occupied with occupancy factor 0.99 whereas (A) and (B) are only partly filled (occupancy factors 0.44 and 0.49).

From Saenger, W., Betzel, Ch., Hingerty, B. and Brown, G.M. (1982). Flip-flop hydrogen-bonding in a partially disordered system, *Nature* **296**, 581–583. Reprinted by permission from *Nature*, copyright 1982, Macmillan Magazines Ltd.

by the hydrophobic side chains. The water structure, whilst being far less ordered and intact than in ice might nevertheless exhibit significantly increased hydrogen-bond correlation times and therefore much greater ability to transport protons than normal bulk water. It is also possible that the hydrogen-bond system contains a proportion of the circular and flip-flop type of hydrogen bond since these are more energetically favourable than the standard hydrogen bond as a result of the entropy gained from the near-equivalence of the distribution of hydrogen atoms.

Now turning briefly to ion transport through membranes, in considering the physics involved in such a process, it appears at first sight, that a considerable potential energy barrier must be overcome for an ion to cross a membrane. This is because ions in solution exist in a hydrated state with the dipoles of the bound water molecules aligned so as to reduce the surface electrostatic potential of the ion. A hydrated ion attempting to cross a region of low permittivity (the lipid membrane for example) therefore faces a large potential energy barrier associated with the increase in electrostatic self-energy of the ion which is inversely proportional to the permittivity of the environment of the ion. For an ion to be transported through a channel with uncompensated hydrogen bonding would require a large amount of energy equivalent to the solvation energy of the ion and this would make the transport process energetically unfavourable.

A number of models have been proposed which attempt to resolve this apparent problem and one of the most successful envisages an ordered water system within the membrane channel. In this case the ordered water structure is held in place by hydrogen bonding to five helical polypeptide chains which form the transmembrane ion channel. The continuous cage-like water structure within the channel is considered to provide low-energy ion-binding sites for the ions which hop from site to site through the channel in an essentially solvated state, with hydrogen-bonding contact to four or five water molecules. As a result of the membrane potential, the water dipoles in the channel are all polarized in the direction of the field. Since only a relatively small amount of energy is required to flip the water molecule orientation against that of the membrane field, this model might also provide some insight into the mechanisms responsible for channel gating.

Lateral proton conduction at the membrane–water interface

In the semi-localized hypothesis of energy transduction, protons are considered to be localized at the membrane–water interface and are thought to be transported to the active site of the ATPase complex via protonic conduction pathways along the membrane surface. A number of experimental studies have been made to investigate the existence of such lateral protonic pathways in membranes and lipid monolayers. In recent work, the pH of Langmuir–Blodgett lipid monolayers at the lipid head group–water interface was monitored by incorporating a fluorescent lipid pH probe into the monolayer. By injecting a small amount of concentrated hydrochloric acid into the aqueous phase at one position in the Langmuir–Blodgett trough and observing the fluorescence a few centimetres away from this position, the transit time for the proton pulse to diffuse to the fluorescence

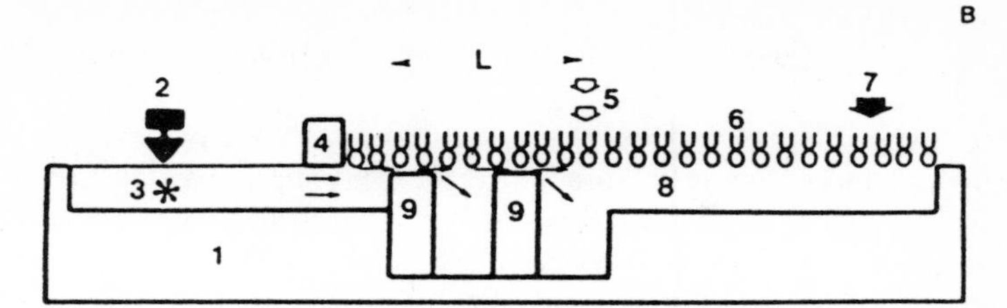

Fig. 5.18 Schematic diagram of the trough used in the lateral proton conduction measurements. 1, trough; 2, injection of acid; 3, stirrer; 4, teflon barrier; 5, fluorescence observation area; 6, lipid monolayer incorporating fluorescent probe; 7, surface pressure transducer; 8, aqueous sub-phase; 9, glass barrier to hinder proton diffusion in the bulk phase and limit it to the window under the film.

From Prats, M., Tocanne, J-F. and Teissié, J. (1985). Lateral proton conduction at a lipid/water interface, *Eur. J. Biochem.* **149**, 663–668.

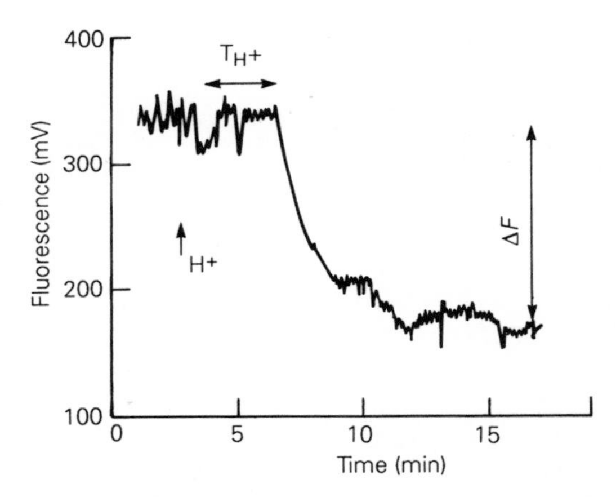

Fig. 5.19 Kinetics of the fluorescence changes (ΔF) of the fluorescein derivatives after an acid jump. At the point marked H^+, acid is injected. T_{H^+} is the delay between acid injection and the beginning of the decrease in fluorescence.

From Prats, M., Tocanne, J-F. and Teissié, J. (1985). *Eur. J. Biochem.* **149**, 663–668.

observation area could be measured, as shown in Fig. 5.18. The diffusion time due to proton migration through the bulk aqueous phase was also monitored by incorporating a fluorescent pH marker into the sub-phase. The lateral diffusion coefficient of protons was found to be some 20 times larger along the monolayer–water interface (diffusion time of 3 min) than in the bulk aqueous phase (diffusion time of 60 min), as shown in Fig. 5.19. By using a value for the proton diffusion coefficient in water of 10^{-4} cm^2/s, the value along the lipid monolayer was then estimated to be 2×10^{-3} cm^2/s. These experiments would seem to provide direct evidence of facilitated proton conduction along lipid–water interfaces.

Further experimental evidence for two-dimensional proton transport at membrane surfaces comes from direct conductance measurements on Langmuir–Blodgett films. It has been observed that gradual compression of fully expanded films of dipalmitoylphosphatidylcholine (DPPC) or dipalmitoylphosphatidylethanolamine (DPPE) produces marked increases in the lateral conductivity of the lipid film below a critical film area. The film packing density at which the onset of enhanced conduction was observed also coincided with an increase in the surface potential and total dipole moment of the film suggesting substantial molecular ordering. It has been proposed that the increased conductance was due to facile proton transport along two-dimensional arrays of water molecules hydrogen-bonded to the lipid head groups. Such a hydrogen-bonded network would be expected to increase in size with increasing packing and would provide interconnecting proton pathways.

Dielectric studies on halobacteria

Halophilic bacteria are rod-shaped organisms requiring a growth medium containing 3–5 M NaCl both for growth and for maintenance of their structure. These cells contain high concentrations of potassium ions and low concentrations of sodium ions with a selectivity $[(K^+_{in}/Na^+_{in})/(K^+_{out}/Na^+_{out})]$ of around 20 000. In the presence of oxygen, many halobacteria synthesize ATP by oxidative phosphorylation. When oxygen is scarce they switch to a photosynthetic mode, using the purple membrane protein bacteriorhodopsin to pump protons across the cell membrane. The proton-motive force generated is then used by the ATP-synthesizing assembly (F_o–F_1) in a spatially separate red-membrane region to phosphorylate ADP.

Recent bioenergetic experiments have suggested that protons produced by halobacterial membranes need not be present in the bulk aqueous phase in order to be utilized in ATP synthesis by the ATPase protein complex. It has been shown, for instance, that there is no change in the pH value of the bulk aqueous phase of illuminated *Halobacterium halobium* as protons pumped by the purple membrane protein bacteriorhodopsin are used by the ATPase system in ATP production. It was concluded that the protons were transported to the ATPase sites along the membrane surface rather than via the bulk aqueous phase.

As we have seen in Chapter 4, dielectric spectroscopy, applied to cell suspensions, can be used to measure membrane and cytoplasmic conductivities and capacitances and to give information about membrane protein mobilities and permeabilities. Dielectric measurements on *Halobacterium halobium* and *Halobacterium marismortui* grown in oxygen-rich conditions have shown two dispersions over the frequency range 1–200 MHz. These can be attributed to Maxwell–Wagner interfacial polarizations associated with the build up of charge at membrane–electrolyte interfaces. Figure 5.20 illustrates the difference in

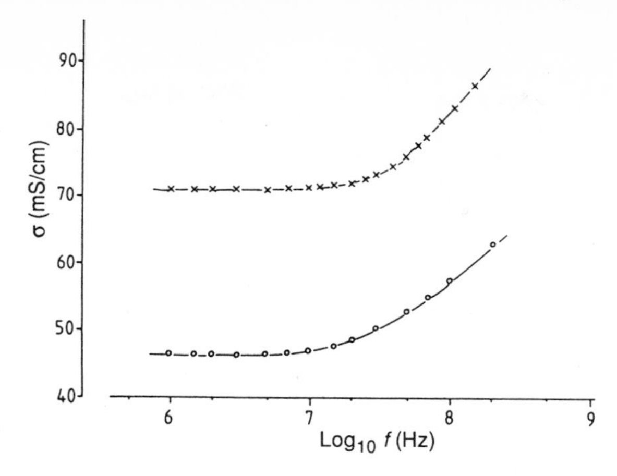

Fig. 5.20 Variation of the conductivity (σ) as a function of frequency (f) for suspensions of the halophilic bacteria *H. halobium* (○) and *H. marismortui* (×). From Morgan, H., Ginzburg, M. and Ginzburg, B.Z. (1987). Dielectric properties of halophilic bacteria *Halobacterium halobium* and *H. marismortui* with reference to the conductivities and permittivities of the cytoplasmic membrane and intracellular phases, *Biochim. Biophys. Acta* **924**, 54–66.

dielectric response of similar volume fractions of *H. halobium* compared with *H. marismortui*. Analysis of these dispersions have shown the membrane conductance of *H. marismortui* to be of the order of 1000 S/cm^2 compared with $\leqslant 1$ S/cm^2 for *H. halobium*. These measurements indicate that *H. marismortui* possesses a plasma membrane which is some four orders of magnitude more conductive than the membranes of other halobacteria and other bacteria in general. The results from permeability studies have also been consistent with this conclusion. The plasma membrane of *H. marismortui* was found to be very permeable to Na^+ and K^+ ions and molecules as high in molecular weight as inulin (MW = 6000) were able to permeate into the cell interior. The membrane of *H. halobium* on the other hand was not permeable to any molecule larger than glycerol. Surprisingly, in spite of its highly conducting cell membrane, *H. marismortui* is able to maintain an internal K^+ concentration 5–10 000 times that of the ambient medium. Even more unusual, this K^+ gradient remains constant even when the cell is starved and the rate of metabolism (as measured by oxygen consumption) is immeasurably small. Recent work has shown that this K^+ gradient is maintained even in the presence of uncouplers which prevent the synthesis of ATP. These observations suggest that K^+ retention in *H. marismortui* is not directly coupled to metabolism. This property is not shared by other halophilic bacteria; *H. halobium* rapidly loses 70–80% of its internal K^+ when metabolism is arrested.

The search for a mechanism to explain these observations has focussed on the unusual properties exhibited by the cell water component in *H. marismortui*.

Differential scanning calorimetry measurements on hydrated pellets from *H. marismortui* have indicated that the 'structured water' component present around proteins from this bacteria may be some three times greater than that expected (compared with bovine serum albumin or lysozyme for example). Recent electrical measurements have made possible an estimation of the K^+ ion mobility within the cytoplasm of *H. marismortui*. A value some ten times less than the mobility of the ions in the equivalent KCl solution was calculated. It seems possible that the mechanism for K^+ retention in *H. marismortui* involves the structuring of a relatively large fraction of cell water as a result of its interaction with specific cytoplasmic halophilic proteins and KCl.

The retention of intracellular K^+ is not the only aspect which is difficult to explain in bacteria possessing a highly-conducting membrane. Perhaps even more difficult is the way in which *H. marismortui* performs phosphorylation. The chemiosmotic theory which, in the case of bacteria, requires the establishment of a bulk mobile proton gradient across the cell membrane is now widely accepted as the mechanism responsible for both oxidative phosphorylation and photophosphorylation. In *H. marismortui*, however, such a gradient would appear to be difficult to establish and has not been found experimentally. One possibility is that the proton gradient might involve localized protonmotive proteinaceous complexes associated with the cell membrane. This scheme would seem to be supported by measurements which have shown that ATP levels are maintained at normal levels despite the permeability of the cell membrane. A recent development of this work has been the observation that the dielectric properties of bacterial cells appear to be sensitive to the bioenergetic state of the cells and to treatment of the cells with metabolic uncouplers and inhibitors.

Selected reading

Bone, S. (1987). Time domain reflectometry studies of water binding and structural flexibility in chymotrypsin, *Biochim. Biophys. Acta* **916**, 128–134.

Bone, S., Gascoyne, P.R.C. and Pethig, R. (1977). Dielectric properties of hydrated proteins at 9.9 GHz, *J.C.S. Faraday Trans. I* **73**, 1605–1611.

Careri, G., Gratton, E., Yang, P-H. and Rupley, J.A. (1980). Correlation of IR spectroscopic, heat capacity, diamagnetic susceptibility and enzymatic measurements on lysozyme powder, *Nature* **284**, 572–573.

Cooke, R. and Kuntz, I.D. (1974). The properties of water in biological systems, *Ann. Rev. Biophys. Bioeng.* **3**, 95–126.

Eden, J., Gascoyne, P.R.C. and Pethig, R. (1980). Dielectric and electrical properties of hydrated bovine serum albumin, *J.C.S. Faraday Trans. I* **76**, 426–434.

Finney, J.L. (1979). The organisation and function of water in protein crystals, in *Water – A Comprehensive Treatise*, Ed. Franks, F. New York, Plenum Press.

Grant, E.H. (1982). The dielectric method of investigating bound water in biological material: an appraisal of the technique, *Bioelectromagnetics* **3**, 17–24.

Kent, M. and Meyer, W. (1984). Complex permittivity spectra of protein powders as a function of temperature and hydration, *J. Phys. D. (Appl. Phys.)* **17**, 1687–1698.

Nagle, J.F. and Tristram-Nagle, S. (1983). Hydrogen bonded chain mechanism for proton conduction and proton pumping, *J. Membrane Biol.* **74**, 1–14.

Pethig, R. (1986). Dielectric studies of proton transport in proteins, *Ferroelectrics* **86**, 31–39.

Poole, P.L. and Finney, J.L. (1983). Sequential hydration of a dry globular protein, *Biopolymers* **22**(1), 255–260.

Rickey Welch, G. (ed.) (1986). *The Fluctuating Enzyme*. New York, John Wiley.

Rupley, J.A., Gratton, E. and Careri, G. (1983). Water and globular proteins, *Trends Biochem. Sci.* **8**(1), 18–22.

Saenger, W. (1979). Circular hydrogen bonds, *Nature* **279**, 343–344.

Teissié, J., Prats, M., Soucaille, P. and Tocanne, J.F. (1985). Evidence for conduction of protons along the interface between water and a polar liquid monolayer, *Proc. Natl. Acad. Sci. USA* **82**, 3217–3221.

Chapter 6

Electron and Proton Transport in Biochemical and Artificial Systems

As was pointed out in Chapter 1, both electron and proton transport in biochemical systems have been the subject of detailed study over recent decades. At present, the basic organization of the electron-transporting pathways (in terms of the sequence of macromolecular carriers) has been mapped out for a number of systems. An example of such a pathway (the mitochondrial electron-transport chain) was given in Chapter 2. Work is now proceeding on the detailed structure of the supramolecular complexes formed by the association of the carriers in biological membranes. At the same time, research is in progress on the mechanisms by which the carriers interact with one another and with the molecules in their environment – both within the membrane and in the aqueous compartments into which many of the carrier proteins protrude. Finally, and from the biochemist's point of view, most importantly, much effort is at present being devoted to unravelling the molecular mechanisms that result in the synthesis of ATP (the 'energy currency' of the cell) at the expense of redox potential energy. It is widely believed that the process utilizes as an intermediate an electrochemical gradient of protons across the membrane in which the electron-transport proteins are embedded (but see Chapter 2 for some recent dissent from this view). The precise molecular mechanism of these energy-interconverting processes, which lie at the heart of cellular metabolism, has been the subject of intensive study since the link between electron transport and ATP synthesis was first recognized.

Structure of electron carriers

The carriers of the electron-transporting chains are, as we have seen in Chapter 2, mainly proteins containing a prosthetic group which is electroactive. Typically, this group contains a metal atom (Fe, Cu) which can undergo redox transitions. The proteins involved in electron-transport represent particularly awkward subjects for high-resolution structural analysis, since they are in general very difficult to crystallize by the usual techniques as they are naturally part of multi-subunit structures embedded in phospholipid membranes. X-ray crystallography cannot readily be applied in these circumstances, although very encouraging progress has been made in the co-crystallisation of membrane proteins with low-molecular-weight detergents thanks to the work of Michel and colleagues. Also, much structural detail is emerging from gene-sequencing studies, followed by amino acid sequence prediction for the gene products and application of the increasingly sophisticated three-dimensional structure prediction algorithms which are now available once the amino acid sequence is known. While the results of such predictive techniques should not be treated uncritically, they frequently provide complementary evidence where low-resolution three-dimensional structures have been identified.

SIMPLE CYTOCHROMES

Although most electron-transport proteins are to some extent embedded within membranes, a number are only loosely associated with lipid structures and do not contain large external hydrophobic domains. Cytochrome *c* is a particularly good example of this type of protein. It is readily crystallized and its three-dimensional structure has been determined at the 1–2 Å level of resolution including subtle differences between the structures of the reduced (ferro) and the oxidized (ferri) forms. A group of related proteins is distributed throughout the animal, plant and prokaryotic kingdoms. A common feature shown by these proteins is their covalently-bound haem group, which is buried in a hydrophobic environment in the interior of the molecule which itself presents a hydrophilic 'face' to the environment. The sixth ligand to the iron (which is in a low-spin state) in this family of proteins is a methionine residue and this factor influences the redox potential which is in the range 100–450 mV (vs. NHE). The *c*-type cytochromes are all basic proteins with many lysine residues exposed to the surface responsible for a net positive charge on the molecule. The charges are concentrated around the edge of the haem crevice and a number of these residues have been implicated in the interaction of cytochrome *c* with its redox reaction partners.

A contrasting group of cytochromes, which has also been studied structurally at high resolution, is the b_5 group. In this family, the haem moiety is attached non-covalently in a crevice which, by comparison with the *c*-type cytochromes is more exposed to the hydrophilic environment. In some members of the family, however, there is a hydrophobic C-terminal domain containing 30–40 hydrophobic amino acids. This feature is responsible for the binding of this protein to microsomal and outer-mitochondrial membranes. The hydrophobicity of the whole protein has so far precluded its crystallization, but trypsin digestion yields the hydrophilic (haem binding) segment which has been crystallized and characterized at a resolution of 2 Å. In this case, a number of highly acidic groups are found on the surface in the vicinity of the haem crevice, giving the protein a net negative charge, in contrast to the positive charge of the *c* group.

The *c* and b_5 cytochrome families contain relatively short (at most 10–14 residues) lengths of secondary structure (α-helix and β-sheet). In contrast the soluble four-helical family (cytochrome b_{562}, *c*′etc.) contain four long (20 residues) lengths of α-helix. The haem group in this class of cytochrome is relatively exposed as can be seen from the 2.5 Å resolution model which has been produced.

CYTOCHROME OXIDASE

A more complicated example of a biological electron-transferring molecular complex is provided by the enzyme which in the living cell is responsible for the oxidation of cytochrome *c*. In oxidizing cytochrome *c* it passes electrons on to the 'ultimate' biological electron acceptor – oxygen. The cytochrome oxidase complex which is found in mitochrondrial membranes contains two haem groups (known as haems *a* and a_3) and two copper atoms (Cu_A and Cu_B) all of which can take part in redox reactions. Its complexity arises not only from the existence of four redox centres, but also from the fact that these centres are not bound to a single polypeptide chain as in the simple cytochromes but to at least seven (and possibly nine) protein subunits with molecular weights ranging from 3400 to 35 300. Chemical labelling of the complex, together with hydrophobicity indices allows the orientation of the subunits within the complex to be inferred. Subunits 1–4 have membranous segments, subunits 2, 3, 5, 6 and 7 all have domains accessible from the cytoplasmic side of the membrane while subunit 4 has domains which can be labelled from the opposite (matrix) side of the membrane. The indications are that the molecule is asymmetric with respect to the membrane with 50% of the molecular mass protruding some 60 Å into the cytoplasmic aqueous phase, where cytochrome *c* binding takes place. Two-dimensional crystallization and electron microscopy confirms these features as well as indicating that the smaller (matrix) aqueous domain is a bifurcated structure protruding some 15 Å into the aqueous phase. The enzyme quite possibly exists as a dimer (2 × 7–9 subunits; 8 redox centres) in the mitochondrial membrane, although some bacterial membrane enzymes which catalyze the same reaction appear to have a much simpler structure. The haem *a* and Cu_A centres are located relatively close to the cytoplasmic (cytochrome *c* binding) surface of the complex, while haem a_3 and the Cu_B centre are in the hydrophobic phase, and closely associated with one another.

PHOTOSYNTHETIC REACTION CENTRES

The reaction centres of the photosynthetic membranes are molecular transducers which convert the light energy of photons into chemically stored energy (in the form of low redox potential components) via a charge-separation process. Such centres exist in higher plant chloroplast membranes and in the membranes of the photosynthetic bacteria. These latter have proved to be particularly useful sources of material for structural studies, especially with the advent of cloning technology, leading to predicted structures from gene sequencing. Furthermore, recent crystallization of the reaction centres of the photosynthetic bacterium *Rhodopseudomonas viridis* has allowed high-resolution three-dimensional structure determination by Deisenhofer and colleagues.

The reaction centres prepared from photosynthetic bacteria are hydrophobic protein complexes containing in addition a number of low-molecular-weight chromophore molecules which are capable of photochemical and chemical redox reactions. For example reaction centre complexes prepared from *Rhodopseudomonas viridis* membranes contain four subunits: a multihaem cytochrome subunit of molecular weight 38 000 and three further subunits designated H (35 000), M (28 000) and L (24 000). In addition to the four haem groups in the cytochrome subunit, the reaction centre contains the following chromophore molecules: four molecules of bacteriochlorophyll *b*, two molecules of bacteriophytin *b*, one non-haem iron and one menaquinone. The L and M subunits have been shown to each contain five 40-Å long helices which can be surmised to make up the membrane-spanning domain. The segments between these helices interact with the cytochrome (probably on the periplasmic side of the membrane) and with the H subunit (on the cytoplasmic side). The subunit has a single long membrane spanning helix, the end of which interacts with the cytochrome. In *R. viridis*, the cytochrome is a firmly-bound component of the reaction centre (due to strong interaction with the C terminal of the H subunit), despite having no membrane

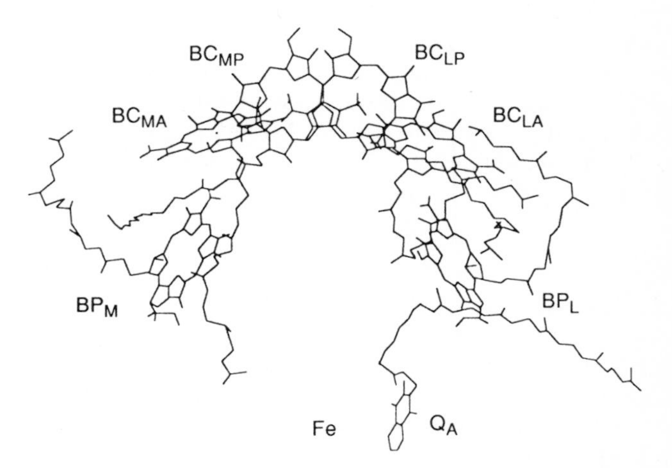

Fig. 6.1 Structural relationship of the pigments in the reaction centre from *R. viridis*. None of the surrounding protein structure is shown.
BC = bacteriochlorophyll; BP = bacteriopheophytin.
Reproduced with permission from Deisenhofer, J. *et al.* (1984). *J. Mol. Biol.* **180**, 385–398.

(hydrophobic) interaction itself. Other species appear to bind the cytochrome more loosely to the reaction centre structure.

Below the four haem groups of the cytochrome and deep within the membrane are the four bacteriochlorophyll and two bacteriophytin molecules symmetrically positioned about the non-haem iron atom. The right-hand structure (two chlorophylls and one pheophytin) is coordinated to the L subunit and a similar left-hand structure is coordinated by the M subunit. The two chlorophylls at the centre of this arrangement are very close to one another with the pyrrole rings in contact. In the right-hand branch of the structure there lies below the bacteriophytin molecule a tightly bound menaquinone molecule, liganded to subunits M and L.

The spectacular progress that has been made with the bacterial photosynthetic reaction centre structures has also benefited workers attempting to grapple with the structures of the PS I and PS II reaction centres (see below) from chloroplasts. These structures have analogous features to those which have been determined at high resolution for the bacterial systems.

A 'stripped-down' version of the structure of the bacterial reaction centre is shown in Fig. 6.1, in which only the carbon skeletons of the pigments are shown, with the protein structure (which is the scaffolding holding this structure together) suppressed from the picture.

Electron transfer within and between redox complexes

The complexes whose structures have been described above participate in electron-transfer sequences and the kinetics of electron transfer both within and between these protein complexes will depend on factors such as distances between redox centres, the redox potential of the species under consideration (although this can be difficult to determine), the mobility of the complexes within the membrane and that of low-molecular-weight carriers. We will now examine recent work that sheds some light on the relative importance of these factors in determining the functioning of some of the complexes described above.

ELECTRON TRANSFER WITHIN COMPLEXES

Photosynthetic reaction centres

The relatively rigid conformation in which the redox centres are held in proteins such as the bacterial reaction centres removes discussions of mobility and diffusion from considerations of their function. The electron moves essentially through a solid-state material of heterogeneous but precisely-defined composition. De Vault has characterized these movements on the basis of the temperature independence of their rate constants as electron-tunnelling processes. Before the structure of the reaction centres had been elucidated (a very recent development), much information on the likely relative location of redox centres had nevertheless been built up from spectroscopic studies on the kinetics of the redox reactions. Since electron transfer within the reaction centres is very rapid, spectroscopic methods with very high time resolution (down to < 1 ps) have been developed. Typically in these experiments, a laser flash of very short duration is used to excite the primary electron donor of the reaction centre and at variable delay periods (in the 1–100 ps range) the adsorption spectrum of the preparation is sampled. Recent advances have allowed sub-picosecond measurements to be made. These studies have led to the following proposed electron-transfer route:

$$\mathrm{PBHQ} \xrightarrow{h\nu} \mathrm{P^{*}BHQ} \xrightarrow{1\ \mathrm{ps}} \mathrm{P^{+}B^{-}HQ} \xrightarrow{4-10\ \mathrm{ps}} \mathrm{P^{+}BH^{-}Q} \xrightarrow{150-250\ \mathrm{ps}} \mathrm{P^{+}BHQ^{-}}$$

where P represents the primary donor (the 'special pair' – P^* in its excited state), B represents a bacteriochlorophyll, H a bacteriopheophytin and Q the menaquinone.

From its optical and electron spin resonance (e.s.r.) properties, P was identified as a 'special pair' of strongly-interacting bacteriochlorophyll molecules. The complementary nature of the kinetic and structural approaches is well illustrated here. The spectroscopic results could be interpreted in different ways, sometimes casting doubt on the existence of the 'special pair'. The resolution of the pigment structures in the reaction centres (see above) clearly shows a close association between two bacteriochlorophyll molecules, supporting the 'special pair' interpretation of the spectroscopic data.

The oxidized primary donor (P^+) is re-reduced by the closely associated cytochrome in *Rhodopseudomonas* reaction centres. In green plants, there are two types of reaction centre, PS I and PS II. In PS II, P^+ is reduced by electrons extracted from water – leading to the evolution of oxygen, whereas the P^+ of PS I is reduced by electrons which originate from PS II, but have passed through the hands of a sequence of electron carriers in the photosynthetic electron-transport chain.

We may now correlate the structural and spectroscopic data, as has been done by Deisenhofer, in an illustrative diagram (Fig. 6.2). Q_b represents a mobile quinone molecule and the $Q_a \rightarrow Q_b$ transfer represents a transfer to a species external to the complex and therefore depends on the diffusion rates of Q_b and the reaction centre complex within the plane of the membrane.

The above sequence is an obvious one to infer from the structure as shown in Fig. 6.1. However, one of the surprising results from recent experiments in this area is the apparent lack of evidence for a role for 'B', the 'accessory' bacteriochlorophyll. The apparent redundancy of the left-hand branch of the pathway illustrated in Fig. 6.2 (no final acceptors present) remains a mystery, as does the precise mechanism by which back reactions are so strongly kinetically suppressed in the systems, leading to highly efficient and rapid charge separation. Much theoretical and experimental work is presently being undertaken to try and construct a rigorous model for charge transfer through this complex.

Cytochrome oxidase

The kinetics of electron transfer between the four redox centres within cytochrome *c* oxidase have also been the subject of intense study by spectroscopic methods, but no method of initiating electron transfer on very short time scales is available by comparison with the photosynthetic reaction centres which can be stimulated with

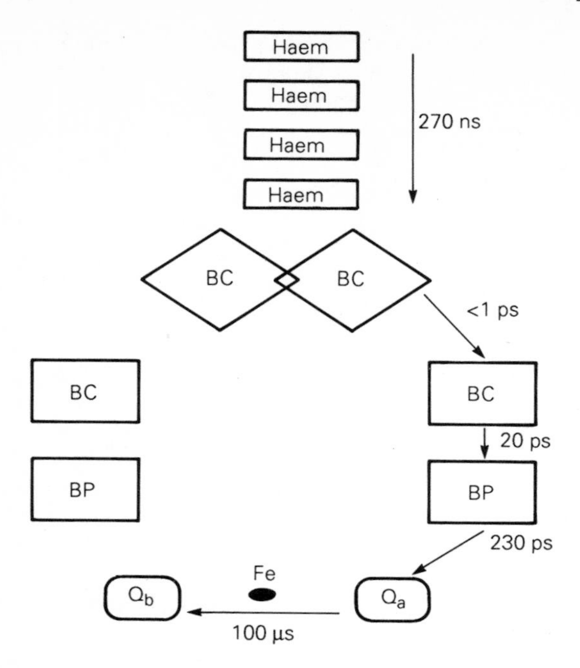

Fig. 6.2 Redox centres in PS I. Electron transfer times between the components of the reaction centre from *R. viridis* are overlaid on a scheme of the structure.

picosecond laser pulses. The usual stopped-flow kinetic methods are available which can be used to pulse the isolated complex with either oxygen (the electron acceptor) or cytochrome *c* (the electron donor) on a time scale of milliseconds, and time scales can be shortened further for the study of the oxygen reaction by the 'triple trapping' technique of Chance, in which the reduced cytochrome oxidase is first saturated with carbon monoxide prior to freezing and photolysis of the Fe–CO bond which allows the oxygen reduction reaction to proceed. This technique has allowed the study of the intermediates involved in the oxygen reduction reaction.

Rapid mixing of the complex with cytochrome *c* has revealed an initial fast step ($k = 8 \times 10^6\,\mathrm{M}^{-1}\,\mathrm{s}^{-1}$). This can be shown to be a transfer to the haem $a/\mathrm{Cu_A}$ redox centre within the complex with transfer to the $a_3/\mathrm{Cu_B}$ centre taking place on a much longer time scale. It appears that although only a single electron is transferred in this rapid step, haem *a* and $\mathrm{Cu_A}$ are both left partially reduced (ca. 50% each) as a result. A maximal overall electron transfer rate of $300\,\mathrm{s}^{-1}$ has recently been quoted for cytochrome oxidase. Recently it has become apparent that the rate of electron transfer between haem *a* and a_3 is relatively slow only when the enzyme is in the so-called 'resting' state. An oxygen-pulsed or partially reduced state of the enzyme is said to have a catalytic activity 4–5 times that of the

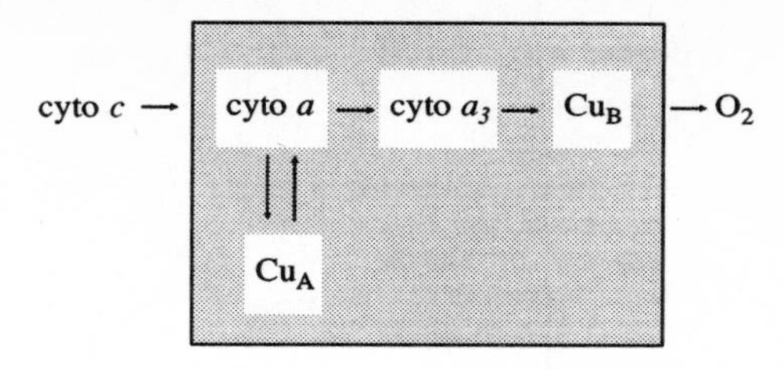

Fig. 6.3 Passage of electrons through the mitochondrial cytochrome c oxidase complex. Four electrons are sequentially passed from cytochrome c through the complex to reduce one molecule of oxygen to water.
After Wikstrom.

resting state, suggesting that considerable conformational changes can take place on partial reduction of the enzyme complex.

The passage of electrons from the cytochrome a_3 moiety to oxygen has been the subject of extensive investigation, since the reaction is particularly complex, involving the transfer of four electrons to the dioxygen molecule. Intermediate oxygen reduction products (peroxide and superoxide) are toxic to many biological processes and are kept to a minimum thanks to the ability of the haem a_3/Cu_B centre to bind oxygen and its intermediate reduction products tightly until the harmless and fully reduced form (water) is released.

The overall passage of electrons through the complex can probably best be illustrated by the schematic sequence presented in Fig. 6.3. In this scheme, Cu_A is not an obligatory intermediate in the electron-transfer sequence. What is interesting about this scheme is that it appears to be the internal ($a \rightarrow a_3$) electron-transfer rate which is rate limiting for the whole sequence and we are therefore forced to conclude that the mere fact that centres of appropriate potential are held within close proximity to one another (in this case ca. 10 Å) does not guarantee rapid rates of electron transfer, as occurs in the photosynthetic reaction centres discussed above. Clearly the polypeptide structure which surrounds the centres can either facilitate rapid electron transfer or can provide a barrier to the process.

The cytochrome bc$_1$ *complex*

This is a family of complexes which exist in mitochondrial, chloroplast and bacterial electron-transport chains and serve as electron carriers between quinols and cytochrome c. Electron transfers within the complex have been the subject of much controversy as the original concept of a linear pathway (Fe/S centre$\rightarrow$cytochrome $b\rightarrow$cytochrome c_1) has had to be discarded in favour of a more complex pathway, involving a bound quinone electron carrier. It is now clear that when two electrons are donated into this complex by the free quinol, one of these is donated to the Fe/S centre, whence a linear chain carries it through to cytochrome c_1 and eventually c, while the other electron reduces the cytochrome b, which is itself oxidized by a bound quinone (or semiquinone) which can eventually again donate the electron to the Fe/S centre. This complex mechanism has

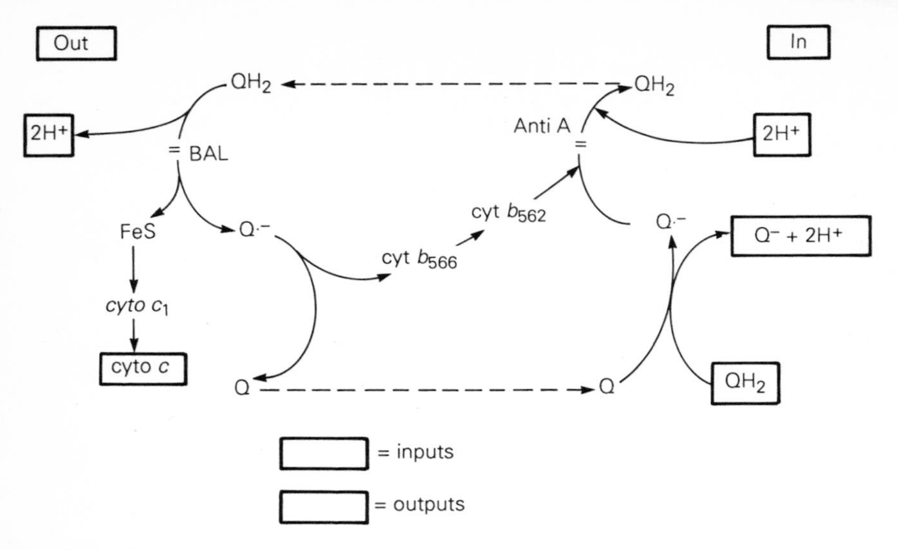

Fig. 6.4 The 'Q cycle' operation of the cytochrome *bc* complex. ——, Reactions involving $H^+ + e^-$ transfer; ---, electron transfers.

presumably evolved partly to cope with the change in the capacity of the electron-transport chain which occurs here (from quinol, a two-electron carrier to cytochrome *c*, a one-electron carrier) and partly to provide a mechanism for the required proton translocation across the membrane (see below). The electron-transfer pathway may be described by Fig. 6.4. For every two electrons which enter the complex from a molecule of QH_2, one is transferred to the Fe/S centre and cytochrome *c*, the other passes through the *b* cytochromes and reduces Q, reforming QH_2. One H^+ is translocated per electron passed out to cytochrome *c*.

Rich has given values for the maximal turnover number of the bc_1 complexes in the range 400–4000 s^{-1}. Again, we are clearly dealing with rates of electron transfer between centres that are well below those found for the primary (photochemically-induced) electron-transfer rates discussed above.

ELECTRON TRANSFER BETWEEN COMPLEXES

For the operation of whole electron-transfer pathways, mechanisms must clearly exist for electron transfer between the complexes which we have described above. To some extent of course, these complexes might represent artefactually small units of molecular organization, which are formed during the necessarily disruptive procedures used in their isolation from living cells. The possibility cannot be excluded that some of the components which we treat as independent units are in fact part of even larger supramolecular structures *in vivo*.

Where diffusion-limited collisions must take place between complexes before electron transfer can take place, it is possible to calculate the effect of parameters such as diffusion coefficient, concentration of diffusing species, etc. on the rates of electron transfer in the pathway. Rich has pointed out that recent studies of this type suggest that plastoquinone mobility might be inadequate to account for the rates of passage of electrons between PS I and PS II in chloroplast membranes. The possibility that in at least some biological electron-transferring systems the molecular organization is more akin to a solid-state model, than the traditional 'diffusion of icebergs' in the phospholipid 'sea' is particularly relevant to possible applications of bioelectronics, since the biological systems might be susceptible to analysis through the well-developed formalism of solid-state theory. In many instances, however, diffusion of proteins, protein complexes and low-molecular-weight intermediates are known to influence electron-transfer processes, and two cases of this phenomenon will now be considered.

The quinone pool in mitochondrial electron-transport chains

Although, as mentioned above, quinones do not always and everywhere act in biological membranes as rapidly diffusing electron carriers between protein complexes, in many instances they do indeed fulfil such a role. Ragan and Cottingham have summarized evidence for what they call 'Q pool behaviour' in a number of different membrane electron-transport systems. Further, they have developed a kinetic model for Q pool behaviour which allows the testing of kinetic data for various systems. They have shown that often systems which appear at first sight not to conform to a Q pool behaviour can be seen to do so if relatively minor adjustments are made to some difficult to measure kinetic parameters. In ideal Q pool behaviour a quinone molecule can interact with large numbers of potential electron donors and acceptors on a time scale consistent with known electron-transport activity through the system. Using data supplied by Zhu *et al.* and a random-walk diffusion model, Ragan and Cottingham estimated that a quinone or quinol might encounter 70 molecules of bc_1 complex (quinol oxidase activity), a similar number of succinic dehydrogenase complexes and half of that number of NADH dehydrogenase complexes (both with quinone reductase activity) within its expected lifetime calculated on the basis of known electron-transport chain activity.

These values suggest that the quinone–quinol system could with ease fulfil the role of a mobile electron-transfer chain. However, it should be noted that the collision frequencies calculated above are critically dependent on the value assumed for D, the quinol diffusion coefficient, which was taken to be $10^{-8}\ cm^2\ s^{-1}$. This value however, is the subject of some controversy and could be an order of magnitude too high. It should also be noted that Q pool kinetics can be mimicked by diffusion of whole protein complexes leading to productive (electron-transferring) collisions, provided again that diffusion coefficients for these complexes are high enough.

The whole area of mobility of proteins and complexes within the membrane as a prerequisite for electron transfer is under intense study and is benefiting from increasingly sophisticated modelling techniques. These studies are of particular importance to models of electron transfer in chloroplast (thylakoid) membranes, where the two photosystems, PS I and PS II, are localized in different regions of the same membrane. There is in this case an obvious need for a mobile carrier of electrons which is able to shuttle electrons between the complexes. Again models suggest that such a role could be fulfilled by plastoquinone and plastocyanin. A high concentration of quinones in the various membrane systems assist in fulfilling this function.

Cytochrome c *interactions with other redox proteins*

The association of cytochrome *c* with the membrane and complexes contained in it has long been acknowledged to be weak, and dissociation with moderate-to-high salt concentrations can readily be achieved. This suggests that ionic interactions are important for the binding of cytochrome *c* to its redox partners, and by implication for its electron-transfer function. Salemme has put forward a model in which cytochrome *c* is free to diffuse in two dimensions on the surface of the membrane, with relatively weak electrostatic interaction with the phospholipids, but strong interactions with the surface-exposed areas of its redox partners.

Studies of reaction kinetics with derivatives of cytochrome *c* modified at single amino acid residues show that modification of certain lysine residues can reduce electron-transfer rates between cytochrome *c* and cytochrome oxidase by 4–7 fold. These lysines are known to occupy positions around the exposed edge of the haem. The reaction rate of cytochrome *c* with its oxidase is essentially diffusion controlled in isolated systems with the large dipole moment of cytochrome *c* (ca. 325 Debye) presumably acting to ensure correct orientation. As mentioned above, once cytochrome *c* and cytochrome oxidase have collided successfully, an extremely fast initial 'burst' of electron transfer can be measured as reduction of the haem a/Cu_A centre of the oxidase.

Cytochrome *c* is also known to undergo redox reactions with cytochrome b_5 with both molecules in free solution, although the physiological significance of this reaction (if any) is not known. Again electrostatic interactions between ionic groups appear to be of great importance in the formation of active complexes. Recent nuclear magnetic resonance evidence suggests the involvement of the cytochrome *c* lysines in interaction with glutamate and aspartate residues of cytochrome b_5. This lends support to the model proposed by Salemme for the interactions of cytochromes *c* and b_5 in which following electrostatic interactions of the kind described above, the haem groups of the two cytochromes are brought into parallel orientation (8 Å apart) enabling delocalized π-orbital electrons from one haem to tunnel through to occupy similar orbitals in the second haem. This process is further facilitated by exclusion of bulk water from the interaction region, resulting in a de-shielding effect on the haems.

Proton movements associated with electron transfers

As was noted in Chapter 1, electron transfer in many electron-transporting chains in a variety of 'bioenergetic' membranes are thought to be mechanistically linked to proton movements across these membranes. Indeed, the study of these proton movements, the energy stored by the resulting electrochemical proton gradient and the subsequent utilization of this energy for ATP synthesis (the cell's energy 'currency') has provided the framework for much research on electron-transferring proteins. As the components of the electron-transport chains have been studied in ever greater detail, it has become possible to propose specific models to explain how the exergonic electron-transfer reactions can be linked to endergonic proton transport across membranes. Wikstrom has shown that cytochrome oxidase activity results in the uptake of $2H^+$ from the matrix of the mitochondrion for each electron transported by the complex. One of these protons finds its way into the cytoplasmic phase, while the other is a 'substrate' proton for water formation with reduced dioxygen. This evidence supports Wikstrom's earlier assertion that cytochrome oxidase acts as an electron-transport linked proton pump in which conformational changes consequent upon redox reactions change pK values of proton-binding groups in contact with the surface of the membrane. Many studies have been performed on cytochrome oxidase reconstituted into artificial phospholipid vesicles where generation of a proton electrochemical gradient has been demonstrated. Research is presently concentrated on unravelling the molecular mechanism by which the proton movements are coupled to the redox reactions taking place in the cytochrome oxidase. The redox centre associated with the proton translocation reaction appears to be either the haem *a* centre or the closely associated Cu_A moiety (see Fig. 6.3).

A different mode of proton electrochemical gradient has been proposed for the cytochrome bc_1 complex of mitochondrial membranes. Here the movement of ubiquinol across the mitochondrial membrane is involved as a carrier of protons in the so-called Q cycle scheme (see above). For every electron passed through the bc_1 complex, one H^+ is taken up from the matrix side and is ejected on the cytoplasmic side.

The proton movements described above are usually envisaged as occurring from one bulk phase to another. This was Mitchell's original proposal in his chemiosmotic hypothesis. This view, however, is coming under some attack as evidence builds up that the measured bulk-to-bulk proton electrochemical gradient is in fact thermodynamically and kinetically incompetent to drive the ATP synthesis under various conditions when ATP synthesis is known to be maintained. Various forms of 'localized' chemiosmotic mechanisms have recently been proposed in which the relevant electrochemical gradient is held to be localized to within the membrane phase and hence to be inaccessible to bulk measurement. The bulk gradient would be capable of equilibrating with the localized gradient only over a relatively long period of time. A recent exposition of this type of hypothesis, which also accommodates the observation that individual redox complexes might interact with specific ATPase complexes has been given by Westerhoff and colleagues. This area is one which is generating a high volume of experimental and theoretical work and developments are rapid.

Slater has produced a hypothesis on free-energy coupling which emphasizes the role of direct interaction between redox and ATPase complexes, relegating proton gradients whether bulk or localised to a secondary ('buffering') role (see Chapter 2).

Electron transfer in artificial systems

In this section we are concerned with electron transfer between the complexes and proteins described above – the electron carriers of natural systems – and artificial donors/acceptors. Of particular interest as an artificial donor or acceptor is the metal or semiconductor electrode, since its potential can be precisely controlled and currents flowing to and from the electrode can be accurately measured. If the biological electron carriers can exchange electrons with electrodes then the possibility exists that the phenomenon can be utilized in a variety of useful devices and novel energy converting systems.

A vital consideration in attempting to apply electrochemical techniques to biological molecules is the choice and treatment of the electrode material. Although in theory any electroactive species will react at an electrode which is at an appropriate potential, in practice kinetic factors can dramatically limit rates of electron transfer (see Chapter 2). In general, the noble-metal electrode surface, so useful in the study of many relatively simple organic and inorganic molecules, is not compatible with the more complex biomacromolecules, and must frequently be specially treated or modified in order to interact productively with biological electroactive molecules.

ELECTROCHEMISTRY OF SMALL BIOMOLECULES

Although in the preceding section we have concentrated mainly on the electron-transferring proteins, there exists a large group of low-molecular-weight electroactive compounds whose electrochemical behaviour is relatively easy to study. Recent advances in the study of these molecules, including quinones, catecholamines, purines and the vitamin B_{12} group of compounds have been reviewed by Dryhurst. We shall briefly consider here the quinones (whose role in electron-transfer chains has been stressed above) and catecholamines, whose function in the body as hormones makes them of considerable importance.

Quinones

The quinones have been mentioned above as important components of electron-transfer chains. From the point of view of the electrochemist, they present an interesting group of compounds, whose electrochemical properties can be subtly adjusted by varying the substituents on the aromatic ring. Among the naturally-occurring quinones are ones with long aliphatic chain substituents, which act mainly to change the solubility properties of the quinone and methyl, methoxy, amino and hydroxy substituents which can have a marked effect on the electrochemical properties (particularly on the redox potential). The quinones undergo a $(2e^- + 2H^+)$ reduction in aqueous solvents to yield the dihydroquinone or quinol. Although the reaction can under some conditions appear to show simple, reversible behaviour, there is evidence that mechanistically it is a complex process. Clearly, since protons are involved, the redox potential for the process will depend not only on the substituents carried by the quinone but also on the pH of the solution.

Simple quinones such as 1,4-benzoquinone can be reduced at a variety of electrodes to the corresponding hydroquinone (or quinol) in both aqueous and aprotic solvents. In aprotic solvents, reduction proceeds by two separate electron-transfer steps, which are clearly distinguishable in cyclic voltammetry experiments, yielding first a radical anion and subsequently a dianion. The presence of protons, or proton donors, however, complicates the electrochemical reactions considerably. Much controversy still exists regarding the precise mechanism of the electrochemical redox reactions of even the relatively simple quinone compounds in aqueous phases. The biologically relevant quinones, however, which are lipophilic due to their long side-chain substituents have been less intensively studied, but such studies as have been made are consistent with the work on simple quinones, in as much as addition of protons also greatly complicates the relatively simple electroreduction reactions. Takamura *et al.* have studied redox reactions of ubiquinone at gold electrodes and found that the quinone is strongly adsorbed onto the electrode surface by the hydrocarbon chain. They draw analogies between the behaviour of various electron-transporting cofactors at the gold electrode surface to that at the natural membrane surface.

For ubiquinone an $E_{p/2}$ value (the potential at which the value of the current is half of the peak value) has been measured at 0.11 V with respect to the normal hydrogen electrode. This value can be at low as -0.24 V at pH values >9 and as high as 0.36 V at $\leqslant$pH 3. The effect of substituents on $E_{p/2}$ (vs. normal hydrogen electrode) in naturally-occurring quinones is clear when we see values for ubiquinone (ca. 0.1 V), tocopherol (ca. 0.4 V) and rhodoquinone (ca. -0.06 V).

Catecholamines

This large group of compounds which function as both long-range hormones and local neurotransmitters have been widely studied as regards their electrochemical reactions. The stimulus to this study has largely been the promise of electrochemical assay techniques for these substances in biological fluids. The anodic oxidation of these compounds is the best studied reaction and the detailed mechanism of this reaction which involves both a two-electron transfer step and subsequent chemical rearrangement has been established for adrenaline and related compounds. While carbon electrodes have been used for most of this work (and for subsequent sensor development) Sakamoto and colleagues have made a study of the anodic oxidation of catecholamines such as adrenaline and noradrenaline on platinum electrodes modified with surface adsorbed layers of other metals such as bismuth and lead which act as surface catalysts. The oxidation proceeds without the production of strongly adsorbing species which are known to poison bare platinum electrodes. The same workers have applied this type of electrode to anodic glucose oxidation.

Other low-molecular-weight compounds

The electrochemical reactions of purines and pyrimidines – the bases which make up the paired structures of the DNA double helix – can be viewed in a similar light to the quinone electrochemistry discussed above in as much as substituents on the aromatic ring structure are largely responsible for determining the electrochemical properties. Adenine (6-aminopurine), for example, can be electrochemically reduced by a complex mechanism involving six-electron transfers, in which the double bonds of the aromatic ring are reduced and the ring is eventually cleaved. Guanine (2-amino-6-hydroxypurine), however, does not undergo significant electrochemical reduction.

Despite the interesting electrochemical reactions of which the purine and pyrimidine bases are capable, in native DNA no electrode reactions are seen. This is thought to be due to the fact that in the DNA double-helix structure, the purine and pyrimidine bases are internalized and not accessible to a metallic electrode. We shall return frequently to this theme when discussing electroactive proteins.

The electrochemical reactions of the redox cofactor molecules NAD, FAD and FMN have all been studied at various electrodes both in the adsorbed (immobilized) forms and in free solution. NAD is the subject of particular study since it acts as a redox cofactor in at least 300 oxidoreductase enzyme reactions. Since in the course of the reaction, the oxidized or reduced form of the relevant cofactor is depleted, and supplying more cofactor is expensive, electrochemical regeneration of cofactor would appear to be an attractive option in technological applications of enzyme-catalyzed redox processes. It has proved difficult, however to find a system in which NADH transfers electrons efficiently and reversibly to an electrode. In fact, deactivation of the cofactor due to electroreduction followed by radical formation has been described. Possibly work on various types of surface modification of electrodes (see below) will yield systems capable of efficient electron-transfer to these important cofactors.

PROTEIN ELECTROCHEMISTRY

Direct electron transfer between macromolecules and metallic electrodes is not easy to achieve. This is because (i) in the absence of specific modification to the electrode, proteins have only a small chance of approaching the electrode in the correct orientation and (ii) having reached the electrode surface, adsorption can easily take place followed by irreversible structural changes with or without electron transfer. There are two approaches to overcoming these problems: (i) the use of low-molecular-weight mediator molecules to carry electrons from protein to electrode and vice versa and (ii) use of surface-modified electrodes which increase the chance of productive interaction between electrode and protein.

Mediators

The use of low-molecular-weight mediators to transfer electrons between electrodes and electroactive macromolecules has been a technique applied for some time as a method for determining redox potentials of these proteins. This methodology has been described by Dutton and has been widely applied in the characterization of the carriers in the various naturally-occurring electron-transport chains. The mediators used have to be matched to the proteins under study as regards redox potential and tables of suitable mediators have been published allowing a choice to be made. Many of the mediators used are aromatic dyes, such as tetrazolium and anthraquinone derivatives.

The mediator technique has been utilized by Cass and colleagues to carry electrons between a graphite electrode and immobilized glucose oxidase. 1,1'-Dimethylferrocene was used as the mediator which substituted for dioxygen, the usual oxidizing agent in the glucose oxidase reaction. This allowed the system to be used as a stable, amperometric glucose sensor. Ikeda has obtained similar results with the same enzyme electrode but using benzoquinone as the mediator.

Direct electrode-protein electron transfer

Although many attempts have been made to demonstrate electron transfer from bare metal electrodes to electroactive proteins, many of the observed reactions appear to be irreversible and are complicated by adsorption and protein denaturation phenomena. A notable exception is the multi-haem protein cytochrome c_3 which has been shown to react directly and reversibly at mercury electrodes.

A considerable step forward was made by Hill and colleagues when it was found that cytochrome *c* was able to react at a gold electrode which had been surface modified with 4,4'-bipyridyl. A large number of compounds have since been tested as surface-modifying promoters of cytochrome *c*–gold electrode electron transfer and the general principle has been established that cytochrome *c* must bind transiently to the surface modifier through either salt bridge or hydrogen-bond formation. The groups on the surface of the cytochrome *c* molecule responsible for those bonds are the lysine residues which lie in the vicinity of the haem crevice. Thus the same groups which have been shown to be important in determining the reaction of cytochrome *c* with its physiological redox partners (see section above) are also responsible for determining reactivity towards modified electrodes.

Similar modifiers have been used with other proteins such as the bacterial cytochrome c_{551}. Further, other surface modifiers have been devised to promote electrode – protein interactions for proteins such as plastocyanin, which have negatively-charged binding domains in the vicinity of their electroactive groups. Work by Hamer and Hill indicates that a number of proteins, including cytochrome *c* will exchange electrons with unmodified metal oxide electrodes (made, for example, from ruthenium or indium dioxide). The surfaces of these electrodes are thought to be negatively charged, thanks to which the cytochrome *c* will orientate productively.

Albery and Hillman have developed a theoretical approach to modified electrode kinetics in which they have attempted to show how factors such as thickness of the modifying layer and concentration of the electroactive species, etc. influence the location and rates of electron-transfer steps in the redox reactions.

ELECTRON TRANSFER BETWEEN ELECTRODES AND COMPLEX SYSTEMS

Many of the electroactive proteins are, in the living cell, part of complex structures in which they co-exist with other proteins and frequently with phospholipids in membranes (see Chapter 2). Thus there is much interest in achieving artificial systems in which electrodes may be used to transfer electrons to and from whole protein and proteolipid complexes and even whole cells. Again this can be attempted either through the use of small molecule mediators or by direct interaction between the electrode and the electroactive biological system.

Bennetto and his colleagues have been interested in using mediators to exchange electrons between whole bacterial cells and electrodes in order to achieve biofuel cells capable of generating electric currents from metabolic energy conversions. In one system they used phenothiazine dyes to transfer electrons from *E. coli* and *Proteus vulgaris* metabolites to metal and carbon electrodes. These mediators can give rise to very efficient electron transfer and consequently high currents allowing extremely energy-efficient fuel cells to be designed.

The photoreactive complexes, photosystems (PS) PS I and PS II from spinach chloroplast membranes have been adsorbed onto platinized platinum electrodes and photocurrents have been measured in the presence of appropriate electron donors/acceptors. Similarly, bacterial photosynthetic reaction centre complexes have been reconstituted into phospholipid monolayers and deposited onto tin oxide electrode surfaces. Electrical activity in the reaction centres can be monitored by monitoring changes in the system capacitance. While Tiede stresses the use that can be made of this type of preparation to follow charge-transfer reactions within the reaction centre complex, other workers (e.g. Agostino *et al.*) have used similar techniques to study electrode-complex charge transfer, which, as they have shown, is light-induced. Wilkinson *et al.* have described the deposition of proteolipid monolayers containing cytochrome b_5 onto solid electrode support materials, allowing electron-transfer reactions to be modified by the cytochrome b_5 layer. Tien has described studies on a system in which photoreactive proteins are immobilized in a so-called black lipid membrane structure and their electrical activity on illumination can be monitored with suitable electrodes placed in aqueous compartments on either side of the black lipid membrane.

In all such heterogeneous systems containing phospholipid and/or fatty acid, the lipid is usually thought of as an insulating medium in which the conducting (or semiconducting) proteins are dispersed. Due to the extremely thin nature of the lipid layer involved, however, quantum mechanical tunnelling (see Chapter 2) might be expected to take place through this structure and Polymeropolous has shown that such currents (up to nanoamperes in magnitude) can indeed be detected through dry monolayers sandwiched between metallic electrodes. Ionic conductivity through the lipid is of course low – this is the basis of the maintenance of cell membrane potentials – but permeability of lipid bilayers to ions is finite and can be measured. Also receiving some interest are the properties of pure phospholipid layers as catalytic surfaces on metallic electrodes for the catalysis of electrode reactions of certain metal ions.

Selected reading

Deisenhofer, J. (1985). The structural basis of photosynthetic light reactions in bacteria. *Trends Biochem. Sci.* **10**, 243–248.

Dryhurst, G., Kadish, K.M., Scheller, F. and Renneberg, R. (1982). *Biological Electrochemistry*. London, Academic Press.

Dryhurst, G. (1977). *Electrochemistry of Biological Molecules*. London, Academic Press.
Frew, J.E. and Hill, H.A.O. (1987). Electron-transfer biosensors. *Phil. Trans. R. Soc. Lond. B* **316**, 95–106.
Friesner, R.A. and Won, Y. (1989). Spectroscopy and electron transfer dynamics of the bacterial photosynthetic reaction centre. *Biochim. Biophys. Acta* **977**, 99–122.
Matthews, F.S. (1985). The structure, function and evolution of cytochromes. *Prog. Biophys. Molec. Biol.* **45**, 1–56.
Rich, P.R. (1984). Electron and proton transfers through quinones and cytochrome *bc* complexes. *Biochim. Biophys. Acta* **768**, 53–79.
Wikstrom, M. (1981). Cytochrome oxidase *a* synthesis. London, Academic Press.
Williams, R. J. P. (1988). Proton circuits in biological energy interconversions. *Ann. Rev. Biophys. Chem.* **17**, 71–97.

Index